A SHEARWATER BOOK

Encounters
with
Nature

Encounters
with
Nature

Essays by Paul Shepard

Edited by
Florence R. Shepard

ISLAND PRESS / Shearwater Books
Washington, D.C. / Covelo, California

A Shearwater Book published by Island Press

Copyright © 1999 by Florence R. Shepard

Library of Congress Cataloging-in-Publication Data
Shepard, Paul, 1925–1996
      Encounters with nature : essays / by Paul Shepard ; edited by
   Florence R. Shepard ; introduction by David Petersen.
        p.   cm.
      Includes bibliographical references and index.
      ISBN 1–55963–529–0 (cloth)
      1. Nature.   I. Shepard, Florence R.   II. Title.
   QH81.S572   1999   99–33862
   508—dc21   CIP

Printed on recycled, acid-free paper

Manufactured in the United States of America
10 9 8 7 6 5 4 3 2 1

McElway, the team leader and world class paleontologist, had a situation on his hands which was not exactly fossil bones. It was a mysterious loss of provender from the kitchen tent. Oddly enough, it was just when we were working out primate evolution, from plant eating to skulking after the leavings of lions, that someone began taking the leftover joints and chicken wings out of the only refrigerator. Whoever it was slipped back in the dark after meals to look over the remains and pick through them. McElway called us together but had no sooner described the problem than he began to laugh. "It's only temporary," he said. "Everybody suffers a little regression now and then."

—Carnassia Molar, *Trophism and Attention*

Contents

Preface

In OCTOBER 1994, Paul Shepard (1925–1996) was informed by his doctor that he had metastatic cancer and had only a short time to live. Until that time, being a man filled with great energy, motivation, and interest in life, he had never really considered dying. For several months he hovered, stunned, on the doorstep of death. Then, after a partial recovery, he began compiling notes and references for his last books. He then went back to the desk where he worked steadily until a few weeks before his death.

I WAS PAUL'S PARTNER for the last decade of his seven on this earth. Long before I first met him, I had read his books and used them in my environmental studies seminars at the University of Utah. Some of these books I had ordered directly from him, but I had never met him in person until Valentine's Day, 1985. On his way to Jackson, Wyoming, for a speaking engagement (and, from there, on to India on a Fulbright lectureship), he asked me to meet him for lunch on his stopover at the Salt Lake City airport. I suggested, instead, that he

arrive a day earlier to speak to my graduate seminar where we were
reading his *Nature and Madness* (San Francisco: Sierra Club Books,
1982; reissued with an introduction by C. L. Rawlins by the University of Georgia Press in 1998). One of my graduate students took him
to the airport early the next morning, and I didn't hear from him
again for four months. In June, on a short vacation from academia at
a cabin in Bondurant, Wyoming, with Jason, a grandson who loves to
fish, I was surprised early one morning to see Paul drive up in a cloud
of dust and his little blue Honda. In the next months our friendship
and commitment grew firmly into a partnership. We eventually built
our own cabin in Bondurant where he could fish the streams after a
good day's work at the desk.

I entered the relationship thinking I would learn all about Paul's
research and writing from the author himself, but that was not to be.
He was a quiet, hard-writing man who was not prone to spend time
talking about his work or about himself. I picked up bits and pieces
by listening to his table conversations with dinner guests, whom he
loved to entertain, and by sitting in on his public and class lectures
whenever possible. He was a brilliant and eloquent speaker and a
wonderful teacher and I never tired listening to him. And on the rare
occasions when he was inclined to talk about his work or his life, I
learned to listen carefully, and not interrupt with questions or asides
that might send him back in that quiet zone. He opened all his papers
and books to me as well as resources that had influenced his own
thinking. I read, for the first time, *Man in the Landscape: A Historic View
of the Esthetics of Nature* (originally published by Knopf in 1967 and
reissued with a foreword by Michael Martin McCarthy by Texas
A&M University Press in 1991). As I read his past writings and gave
him feedback on work in progress, I began to get a glimmer of his
brilliance.

During the last months of his life, Paul set about completing his
lifework. He made revisions of two books that were then in the final
stages of publication. *The Others: How Animals Made Us Human*
(Island Press, 1996) he saw as a companion piece and extension of
Thinking Animals: Animals and the Development of Human Intelligence
(originally published by Viking Press in 1978 and reissued with an

introduction by Max Oelschlaeger by the University of Georgia Press in 1998). He also completed his editing of *The Only World We've Got: A Paul Shepard Reader* (Sierra Club Books, 1996). He then set to work on two new projects. The first was to compile certain previously published essays into books; the second, to complete a manuscript on the legacy of our hunter/gatherer forebears that would be a sequel to his first book on this subject, *The Tender Carnivore and the Sacred Game* (originally published by Scribner's in 1973 and reissued with an introduction by George Sessions by the University of Georgia Press in 1998). Paul completed the first book of essays, *Traces of an Omnivore* (1996) with an introduction by Jack Turner, and finished the manuscript for his hunter/gatherer sequel, *Coming Home to the Pleistocene* (1998), edited by myself. Both were published posthumously by Island Press. With the present book, *Encounters with Nature,* Island Press and I carry on our commitment to bring more of Paul Shepard's gift to readers.

In his preface to *Traces of an Omnivore,* Paul wrote that from where he stood at a watershed in his life (at the time he was looking the Grim Reaper in the eye), the future remained as obscure as ever and the patterns of the past were difficult to discern. On second thought, he admitted: "There is some progression . . . that is not imaginary." I profess no expertise in analyzing that progression. Yet in the years since his death, editing his manuscripts and papers, I have become more starkly aware of the vast body and deep insight of his literary legacy. In his essays and books he repeatedly went back to certain themes he had uncovered first in *Man in the Landscape:* the aesthetics and perception of landscape and nature; the antecedents of our ecological thought and practices (as well as our ecological madness); the legacy of our hunter/gatherer forebears; animals and their pervasive influence on our human being, cognition, and imagination; ontogeny, the development of the individual in complicity with nature and culture; and "place" as the grounding of our being. In the last years of his life and following the writing of *The Sacred Paw: The Bear in Nature, Myth, and Literature* (Viking, 1985), he came closer to the bear and its importance in the mythology and cosmology of northern hemisphere peoples. During his life, and in each of his books, he reworked these major themes from different perspectives, digging down to the radical grounding of our perceptions and beliefs. Professor of Human Ecology and Natural Philosophy is the title he chose for his endowed chair at Pitzer College, one of the Claremont Colleges in California, and this title fits perfectly the broad spectrum of his work.

When I began to select and organize the essays for this book, I went first to those Paul had singled out for publication and thought about the way he had put them together. I then went beyond these essays to unpublished pieces. With the help of Barbara Dean, editor at Island Press, and David Petersen, the author of the introduction to this volume, I made the final selection. In this book I have tried to cluster pieces that clarify the progression of his work in two primary areas of interest, animals and place. The two parts are not arbitrary, but neither should they be taken too literally, for each essay exemplifies Paul's unified and interdisciplinary approach to research. One of the hallmarks of his writing—a trait that has, at times, been a disadvantage since it cannot be categorized clearly—is that it synthesizes views from diverse fields.

My hope in publishing these essays is to bring the two topics, animals and place, into sharper focus for readers as well as illuminate the context of the life that brought them into being. The introduction by David Petersen, along with the text of the essays, satisfy the first purpose. My biographical sketches and explanations introducing Part I and Part II, I hope, will convey the context of Paul's full and very complicated scholarly life during the writing of these essays. In this collection the reader will find early essays that presage contemporary environmental thinking as well as some of Paul Shepard's very last thoughts.

My editing of Paul's published essays was minimal: minor changes in punctuation, terminology, and form plus any editorial changes Paul had made to the essays as he compiled them. Although readers should expect to find certain themes reappearing in these essays, I have tried to remove out-and-out repetitions. On the unpublished essays, some of which were first drafts, I did more extensive editing. I did so only for clarification, however, and, as in all my editing of Paul's work, I have refrained from changing or extending his ideas. In these unpublished pieces the language and style are often more relaxed and, I think, an interesting contrast to the more polished essays published previously.

Encounters with Nature has brought me once again into collaboration with Barbara Dean, editor at Island Press, a relationship I continue to cherish as one of the best of my life. We are joined by David Petersen, a passionate and exuberant writer and naturalist, whose "hunter's heart" is very close to Paul's and whose lyrical essay introduces this volume. Barbara's wisdom and insight and Dave's honest feedback in our three-way conversations were invaluable as I selected the essays. I am greatly appreciative to them for their keen abilities, their dedication to Paul's legacy, and their patience with my limitations. My editorial efforts were greatly enhanced by the able hand and insight of Don Yoder, copy editor, and the guidance of Barbara Youngblood, an editor at Island Press. And I shall forever be grateful to two daughters, Lisi Krall and Kathryn Morton, who helped Paul compile the references and materials for his last books.

For Paul's many followers and friends, this volume of essays will bring a greater appreciation of the emergent nature of his work. For new readers, it offers an overview of his writing and his life and will help them to grasp his unique and overarching vision. For all, it should be assurance of his great love for this complex and beautiful planet earth as well as its human and nonhuman constituents.

My life with Paul now often seems dreamlike. At times it seems as tenuous as the first raindrop in a placid pond—an event all the more stunning for its momentary appearance. Although Paul's words continue to bring wisdom and sustenance to me, I sincerely hope that reincarnation is a possibility. We desperately need the spirit of Paul Shepard in our midst.

FLORENCE R. SHEPARD
Salt Lake City, Utah

Introduction

WHILE I'M AMONG THE MOST AVID, I'm hardly the most apt student of Paul Shepard and his "subversive science" of human ecology—the revolutionary study of the overarching significance of humanity's coevolution with wild animals (Shepard's *Others*) in shared wild environs. Thus, while I'm profoundly honored to have been asked to introduce *Encounters with Nature,* this stirring overview of the incremental evolution of an utterly new and infinitely old way of envisioning and addressing life, I'm not so sure I'm equal to the challenge. I am no academic. Consequently, what follows comes as much from the heart as from the head. The most honest thing I can do for Paul, and for you, is to try and relay what he and his revelations—so compellingly revealed in the pages to follow—

have meant in my own life while gently suggesting they can do as much in yours.

⌖

LOOKING BACK TODAY from the vantage of half a century's living, I'm surprised I was able to untangle as much of it as I did for myself—or perhaps I should say in spite of myself—as evidenced by an early instinctive iconoclasm that manifested itself, as the years spun out, in many and various forms, including . . .

An untutored and playful suspicion that the compelling and esoteric biblical story of humanity's fall from grace in the Garden of Eden is in fact a pastoralized metaphorical account of post-Pleistocene humanity's gradual transmogrification, from a socially tribal, regionally nomadic, spiritually animistic hunting/gathering lifeway, to sedentary, nationalistic, messianic, agriculture-based civilization. (Plug in the premise that "Nature is God" and the rest falls right into place.) . . .

An innate passion for wild animals and wild places and the wild-hearted people who haunt those places, manifested in a soulful hunger to be *out there* among them . . .

An unswerving conviction that the acts of hunting, killing, and eating wild creatures, undertaken quietly, reverently, with heartfelt respect and gratitude, are not merely moral and honorable but essentially sacred . . .

And perhaps most significantly, validating the worth of all these "primitive instincts" and others that have guided my life, I long ago recognized that in feeling as I do, by embracing life in such a feral, contrarian, hands-on fashion, I am a "cultural misfit" only because civilized culture so radically mis-fits our wild human nature.

And so it is that, looking back now, I'm surprised and grateful I deciphered enough on my own to provide the determination to live my life—to paraphrase singer-songwriter Iris Dement—the way I want to, not the way I "should." And the way I've always wanted to live is quietly, simply, and wild to the bone.

Yet any logically solid comprehension of what prime mover, if any, motivates these inborn "anachronistic" feelings—hunches, convictions, instincts—and how all the pieces fit together, if they do, continued to evade me, elusive as a deer, lurking deep in the shadow-forest of intellectual ineffability. Then I discovered Paul Shepard, and he explained my life to me.

Certainly there were other significant teachers, exemplars, and mentors before I found Paul Shepard. Edward Paul Abbey—legendary Southwestern writer (*Desert Solitaire, The Monkey Wrench Gang,* and a score more), environmental warrior, and cultural contrarian—was outstanding among them. These freethinkers, Abbey and the others, helped steer me ever closer to a personally meaningful and, I like to think, relatively honorable lifestyle guided by a logically tenable organic philosophy arising from a nature-based, neo-animistic Weltanschauung.

But it would take Paul Shepard to conjoin all the loose ends with his scientifically rigorous and spiritually satisfying vision—speaking, gently teaching, through a synergetic series of books, his prose often painfully academic yet uniquely elegant, subtly humorous and magnetically attractive, as rich and heady as good Irish cream. And while, as I've said, it's mostly my own life Shepard has explained to me, it's *your* life he can help explain to you.

That is: We all share the same DNA wiring diagram, or genome, engineered by evolution over millions of years of trial and error to respond positively, or negatively, to certain ways of living and looking at life. Consequently, we're all far more alike than different. If I am male and live in the country, and you are female and live in the city—so what? We're all in this together and that common genetic taproot runs far deeper than mere race, gender, lifestyle, or, especially, *culture du jour.*

Best of all, no leap of faith is required—no demand that you believe in something you can neither touch with your senses nor rationally comprehend. Paul Shepard's human ecology is not mysticism; it's pragmatism with a bright spiritual edge.

PAUL SHEPARD has been called a prophet by those whose business it is to make such calls. And it's true. His vision comprises not merely the boisterous calamity of civilization but life's basic underlying truths and humanity's deep-time primogenitors, rooted and sustaining.

And driving Shepard's vision was his genius for synthesis. While most scholars, including many of staggering intellect, work diligently within the secure and sturdy walls of conventional academia, viewing the world through the narrow slit-windows of their particular interests and expertise—their own little culturally conjured corners—Paul Shepard moved beyond convention, out under the open sky, where he was free to turn and peer in every direction—up, down, behind, and ahead—getting the big picture, the long view of life; blending, sifting, and culling massive mounds of scientific, historical, and deductive data—drawn from biology, ecology, ethology, anthropology, archaeology, psychology, sociology, philosophy, and even, as we'll see in Part II, art—searching out shards of truth hidden among the dross, then fitting those bits together to provide logical and meaningful shape to the cosmic crossword of life.

And a central piece in that puzzle, perhaps *the* central piece, as Part I makes clear, is the hunt for sustenance, spiritual as well as physical, that the author, employing clever double entendre, calls the Sacred Game.

And this, for me, is where it gets real personal.

ALL MY ADULT LIFE I've asked myself (as others have asked me) one nagging question, over and over, to the point that trying to answer has become an intellectual avocation: Given my all-consuming love for wild animals, why do I hunt and sometimes kill them?

My father's only recreation was work. Thus I had no persistent mentor, nor even an exemplar, to nourish my taste for wild meat, wild country, and wild-ness. Rather, from the beginning, hunting in my life has been a self-generative passion—like a fledgling raven's compulsion to fly, a fish's need to swim, like *instinct*. Whatever its inspiration, hunting has been a blessing in my life, as it has in many another. And Paul Shepard, like no one else, can say precisely why.

To wit: With great diligence Shepard establishes that millions of years of increasingly social and strategic pursuit of increasingly large and savvy wild beasts facilitated the evolution of human intellect, helped to intensify and shape our uniquely primate social self-awareness, and bound us, body and soul, to the rest of wild nature; that this upward-spiraling process of natural selection, genetic shaping, and intellectual growth culminated during the upper Pleistocene, 50,000 to 100,000 years ago; and that these deep-rooted venetic (hunting) instincts remain etched in the human genome today. With all this well established as background, Paul Shepard notes in his 1959 essay, "A Theory of the Value of Hunting":

> In urban and technological situations, hunting continues to put leisure classes in close touch with nature and to provoke the study of natural history and to nourish the idea of conservation.

True certainly for Shepard himself—who hunted and trapped small game as a boy in Missouri, hunted deer and flew hunting hawks as a young man, and fly fished (hydraulic hunting) throughout his life. Similarly, hunting helped ignite the nature love and conservation ethics of such as Aldo Leopold, Ernest Thompson Seton, John James Audubon, Theodore Roosevelt, Olaus Murie, C. Hart Merriam, Henry David Thoreau, Gary Snyder, Dave Foreman, and many more prominent naturalists, conservationists, ecologists, and eco-warriors.

And it's true as well for me. Like all the others, it was my youthful innate yearning for the hunt, with all the richness of outdoor adventure and discovery that true hunting implies, which initially lured me out of town and away from the usual, typically pointless, teenage pursuits and has continued to bring me ever deeper into the natural world, ever deeper into my essential self, ever since.

Moreover, it was my desire to hunt *well* that led me to become a bug-eyed student of animal habits and habitats. Setting out initially only to kill and eat a bit of nature, before I knew it I'd become an amateur naturalist and born-again nature lover. And what we know and love, we will fight for. After this fashion, exactly as predicted by

Paul Shepard way back when, hunting has guided my personal development and inspired my ecological ethic, all for the better.

This is not to suggest that the path from hunter to conservationist is a universal progression, but merely that it happens all the time. In sum, hunting remains among the grandest potential manifestations of the human animal's genetic—thus spiritual—union with wild nature, acquired across thousands of generations of total, passionate immersion in the world of wild animals, wild places, and wild life.

And this ancient, visceral connection to the *rest* of wild nature (as we ourselves are wild and natural), Shepard shows us, is both the essence and culmination of our evolved genetic heritage: the emotional glue binding Others and Place to us and eternity, powering the pragmatic spiritualism of Shepard's human ecology.

And while my old friend Ed Abbey didn't know human ecology by that name, he knew it by heart.

❧

SHEPARD AND ABBEY: As diverse as they were in personality, experience, and voice—and though they never conspired or *even* met (Paul did try once to "track Ed down" while visiting the infamous recluse's hometown, but the phone was unlisted)—these two were yet a pair, twining in my heart and mind.

Which I mention here primarily by way of saying: If you know and loved Abbey but have yet to meet Shepard, get set for a treat that could change your life. To illustrate the "accidental synchronicity" in the thinking and eco-politics of these two apparently disparate intellectual revolutionaries, I offer a brace of examples.

First: Shepard frequently praised, paraphrased, and quoted Lewis Mumford (*The Myth of the Machine* et al.), an insightful and outspoken cultural critic whom Abbey repeatedly "nominated" for the Nobel Peace Prize.

Second: Abbey boldly (like Shepard, boldness was his norm) championed the removal of all private commercial livestock from all public lands in order to restore free-ranging, self-sustaining herds of bison, wapiti, deer, pronghorn, and other native wildings—not only by way of returning a tortured ecology to a more natural, healthful, and aesthetic state, but so that able-bodied Americans could more easily experience the challenge and satisfaction of earning at least a sacramental portion of clean lean meat the good old-fashioned hard way . . . as opposed to the civilized norm of relegating our killing to hard-hearted professionals and buying our meat (cruelly domesticated and bloated with engineered fat and chemicals) at the supermarket, thus reducing sacred flesh to lowly product while disguising and denying our naturally selected role as predatory omnivores.

Nothing could be more Shepardian! By publicly proffering such a "savage" and culturally "subversive" proposal as this, Abbey was calling for a return to what Shepard applauded as "sacramental rather than sacrificial trophism"—that is, the hunter/gatherer/animist's serious, intimate, and ultimately *sacred* approach to killing in order to eat and live . . . as opposed to the agriculturist's meat-market-exchange view—this being but one item on Shepard's long list of "accessible Pleistocene paradigms," those lessons from our deep coevolutionary past with others and place that could, with little effort or cost, improve both our physical and spiritual lives today. And certainly tomorrow.

And so it was, in a multitude of ways, that Abbey was a passionate if inadvertent disciple and proselytizer of the "future-primitive" aspect of Shepard's overall vision. Clearly there was some coevolution going on here as well. But while Ed was a self-professed "scientific lightweight" and "philosophical dilettante," Paul was a tenacious researcher and singularly heavy hitter in both fields.

Which is not to say that Abbey was ignorant of Shepard. Among the many things I admired about Abbey was his ignorance of almost nothing of importance; that was the upside of his self-deprecating "dilettante" nature. Specifically I recall discussing with Ed the astounding explication of the Garden of Eden saga in Shepard's book *Nature and Madness* (1982) and how Shepard's science-based and deep-history-rooted revelations not only fit our own independent, uneducated hunches regarding this premiere enigmatic parable but validated and informed those hunches with astounding insight and scholarship.

＊＊＊

THE SUBTITLE of Paul Shepard's 1996 masterpiece, *The Others,* is *How Animals Made Us Human.* Upping that ante, I can testify that animals have made me human twice: first, in accord with Shepard's careful explications of human phylogeny, as a member of the species *Homo sapiens*; second, as an intractable individual within that grand if befuddled species.

As Shepard suggested across a long and distinguished career of teaching, lecturing, and writing, and Abbey shouted from every canyon rim (both men, each in his own way, echoing Thoreau's deathbed comment to the preacher who was trying to "save" him), I too am convinced that the most logical, rewarding, honest, and honorable course is to focus on just "one life at a time," with as much of that life as possible lived out and about in what remains of the earthly heaven of wild nature—the only heaven we can ever know; the only heaven I ever *want* to know.

More precisely: Any postmortem "paradise" lacking bugling autumn elk, fire-gold aspens rattling in a fresh October breeze, scarlet-throated trout leaping for joy in sparkling mountain water, howling wolves, yipping coyotes, the eerie yodeling of loons, the humbling aliveness that comes with the possibility of meeting a grizzly bear or mountain lion around the next bend along some shadowy forest trail, the gritty ecstasy of love on the rocks, the spicy bite of a good cigar, the smoky bouquet of a cask-aged whiskey—any so-called heaven lacking such earthly blessings as these would be pure hell for me.

And I know Paul Shepard agrees, having remarked so often, in so many ways, that any world, no matter how heavenly, populated solely by humans, no matter how angelic, would be far too lonely a place to bear.

After this stirring fashion, their visions rubbing together from afar, Shepard and Abbey generated a high-voltage synergy that electrified my life and still does: yin and yang incarnate. What brilliant sparks would have flown had those two ever knocked heads across a flaring campfire!

⁓⦵⦵⁓

SADLY, I never got to share a campfire with Paul Shepard. I never even laid eyes on the corporeal man; never heard him lecture to a hard-nosed academic audience or tell a rich joke in private; never felt the warmth and firmness of his handshake; never stood knee-deep in sun-spangled water to watch the graceful unfurling of his fly line against a cobalt western sky. And this, the sum of all those nevers, is a lingering loss in my life.

Yet helping to fill that void is an increasingly warm friendship with the prime mover behind this volume: Paul's wife, widow, and editor, professor emerita Florence Rose Shepard . . . "Philosophy Rose," as one of her two doctor-daughters jokingly introduces her, "the existential phenomenologist and mythopoetic reconceptualist."

Having read most of the much that's been written in celebration of the life and work of Paul Shepard, flowing from a profusion of pre-eminent pens, and not to belittle the many long and close relationships Paul maintained in his active and outgoing life, I remain unaware of anyone who better knew Paul Shepard the man, or who better understands and explicates Paul Shepard the scholar/teacher/prophet, than Florence Shepard.

Encounters with Nature is Paul's triumph, not Flo's. (Her triumph is *Ecotone* [1994].) Yet Flo's invisible editorial touches and intimately informed biographical backdrops help to make this volume among the most accessible and thus the most useful of Paul Shepard's books—an appropriate capstone to a long and luminous, if massively underappreciated, career.

Among the singular strengths of this invaluable book is that its twenty essays span four full decades but are neither dated nor a mere helter-skelter smorgasbord of loosely related writings sandwiched unconvincingly between covers. Rather, these chain-linked essays chronicle an interlocking progression of knowledge, insight, and prescience that codifies Paul Shepard as the father of human ecology—this spanking new yet infinitely old way of approaching and aligning science, philosophy, and the inherent sacredness of all life. Even the earliest of Shepard's scientific revelations and philosophical concerns, reproduced here, have not merely weathered the rigors of decades of academic scrutiny but have become increasingly validated and urgent—testimony to the timeless worth and import of Shepard's grand vision.

⁂

SIGNATURE ASPECTS of creative genius include an eager willingness to abandon the comfort of convention in order to range far afield, searching diligently for insight and epiphany in new and unlikely places; an ability to examine the old and seemingly ordinary from fresh, extra-ordinary prospects; the ursine curiosity and tenacity to leave no stone of possibility unturned. Paul Shepard, more than any other thinker I know—and I know a few—embodied just such courageous contrarian qualities, just such ability and resolve, all of it fired by a burning personal passion.

Restated: While traditional tenured academics are largely in the business of building castles and defending them, Paul Shepard devoted himself to tearing castles down—not to be deconstructive in any mean competitive sense but to clear away the accumulated overburden of centuries of cultural masonry in order to expose the foundations hidden beneath. This constructively deconstructive approach to digging for the deepest secrets of life not only informs human ecology but explains its self-inflicted sobriquet, "the subversive science," in that its nondoctrinaire approach to research and interpretation necessarily pokes and jabs—with the fire-hardened spears of science, logic, and truth—at many of our most sacred, and often flatulent, cultural cows.

⁂

IT WAS YET ANOTHER PAUL—Gauguin, that notorious French painter of bronze-skinned Tahitian beauties—who coined my favorite sum-mary of civilization's most persistently pressing and perplexing questions—that nagging triad of existential/metaphysical queries which five thousand years of East/West philosophy (including what Shepard calls "a dime-store counter of psychological trinket-theories of the self" along with a spiritually stupefying pharmacopoeia of man-made messiahs) have not merely failed to answer but rarely even address:

> Where do we come from?
> What are we?
> Where are we going?

As one long saddled with what Shepard describes as "a distinctive primate preoccupation [with] Self and Society," I have been tormented and entertained for decades by the three-pronged mystery summed up by Gauguin. Across those years, I've inspected a bucketful of approaches to coming to terms with universe, society, and self—while never once feeling moved to embrace any one of them, or even any combination, as The Way.

Then, belatedly but not too late, I met Paul Shepard—"met" him, that is, through his books, which paint an intellectually challenging, academically detailed, spiritually revolutionary, lively and lovely portrait of the inchmeal "creation" of humans as increasingly introspective wildings living and learning among, eating and being eaten by, the Others. And that explication, that portrait, so compellingly conveyed in this very book, substantially answers both where we come from and who we are.

The third of life's Big Mysteries, where we are going, enticingly *remains* a mystery, as no one can see clearly into our collective future or beyond individual extinction. Yet, armed with Shepard's insights, we can at least be certain that the quality of our posterity—Others, Place, People—will in largest part be determined by our allegiance, or continued lack thereof, to our Pleistocene genetic mandate. And that is Paul Shepard's gift to us . . . that is the gift of this book.

As Florence Shepard explains in a closing address, delivered to the Sixth World Wilderness Conference in Bangalore, India, in 1998, Paul Shepard's books comprise:

> a mirror held before us "thinking animals" that reflects our primal human being. This image, if comprehended and lived fully, Paul counseled, can make us at home on Planet Earth, rather than ecological misfits. We recognize this image, for at the heart of our identity is a fundamentally wild being, one who finds in the whole of wild nature all that is true and beautiful in this world.

To fend off speculative criticism in advance: Neither Paul Shepard nor any of his students, myself included, is saying we should abandon all modern technology, conveniences, and comforts and return en masse to hunting wild animals and berries for dinner, dressing in skins and living in caves or tents—nor even, as I may at times seem to be suggesting, that we should all turn to "recreational" hunting as a means of getting in touch with our inner Neandertal. This is not what is meant by following "accessible Pleistocene paradigms," not what is implied by the term "postmodern primitivism," not Shepard's message.

What the study of human ecology does suggest is that (1) since we haven't even begun to evolve a "postmodern genome" adapting our physical, intellectual, social, and spiritual needs to our Johnny-come-lately, increasingly urban and urbane, agricultural, pathologically political, suffocatingly overcrowded, hyperfrenetic, polluted, and pathetically "virtual" existence; and (2) since contemporary culture has staggered so very far off the map of human and ecological needs; then (3) perhaps it's time, for the first time in ten thousand years, to reconnect, at least spiritually, with our wild human nature.

Put otherwise: The "true meaning of life" can neither be credited to nor explained by civilization and its self-justifying otherworldly Mysticism Inc. Rather, from the recent moments of their respective inventions, civilized cultures have been models and promulgators of deviation from the natural norm. And it's exactly that natural norm we must embrace in order to survive, much less to prosper.

Where are we going?

I surely don't know. But with the guidance of such revolutionary, critically important, and intellectually sophisticated de facto "self-help" texts as this, at least and at last we have the tools to seriously begin digging for our true, earthbound roots.

DAVID L. PETERSEN

PART I ▐ Following the Animals

P AUL'S PASSION AND FASCINATION with animals were grounded
in a childhood filled with watching them, being in the presence of
them, and keeping them and continued throughout his lifetime.
From a childhood spent collecting bird eggs and butterflies and following an older boy on his trapline, he progressed, in adolescence, to
falconry, and from there, in adulthood, to becoming an avid bird and
animal watcher, a naturalist, a teacher in environmental studies, and a
writer. In his later life nothing pleased him more than to take a hike
with students, children, or friends. Keenly aware, he picked up on the
multiple nuances and complexities of nature and interpreted them
for his companions to their delight and satisfaction.

Paul, however, also possessed the rational mind of a researcher and spent a good portion of his life in libraries poring over musty texts. He insisted that he had acquired an allergy to book dust—substantiated by the sneezes that identified his whereabouts deep in the stacks. His unique interpretations of nature and ecological perception, human/animal relationships, and human development grew out of years of study. He examined artifacts and popular and ancient iconography in museums and galleries, too, and in contemporary representations in art as well as advertisements and cartoons. Stocked with references from his in-depth research in a multiplicity of fields, he traveled extensively to check on leads uncovered by his research.

Home from his excursions to libraries and trips abroad, he would type energetically for six to eight hours a day. (In his younger days, he said, he could work into the night.) The movement of his hands, like a composer's at the piano, released the stored and categorized thoughts that then surged forth in compelling language. Critics, who considered him too speculative to take seriously, did not fathom the extent of his search or his ability to synthesize diverse perspectives into original scenarios concerning the ecology of our species.

His research did not lead linearly from one topic to another. His tracking was more on the order of animal trails in the woods— woodpaths, Heidegger called them—that disappear suddenly only to reappear unexpectedly and lead us in new directions, the pattern repeating itself over and over. The pervasive themes of his writing returned to be picked up time and again, followed in a different direction, and then reconsidered.

❧

PAUL'S LAST PUBLIC ADDRESS, "The Origin of Metaphor," was delivered in November 1994 shortly after he was diagnosed with metastatic cancer. It was presented at the Museum of Natural History in New York City as part of the "Writings on the Imagination" lecture series sponsored by The Touchstone Center and published posthumously by the *Touchstone Center Journal* (1997). It is a befitting introductory essay to this book, and particularly to this section, "Following the Animals," for it is an ode to the animals he so loved and to whom he owed much of his own insight into their influence on our human understanding and being. After publishing *The Sacred Paw*, Paul allotted the bear a place of honor among the animals. In this essay he explores the etymology of "bear" as well as the importance of the creature itself in the evolution of human development and the emergence of the mind. As he was one of the few individuals in the United States, and in the world, interested in bear mythology and cosmology, the bear appears frequently in his essays. His last research trip was to Scandinavia in 1993 to seek out what remains of ceremonies on the cult of the bear.

Paul did not abandon a question after publishing a book. Each book raised more questions that he would continue researching and reworking, always with the notion of writing another book that would clarify or extend the premises of the first. Following the publication of *Nature and Madness* in 1982, he continued to explore the importance of nature in the ontogenetic development of children, especially with regard to the role of animals. "Animals and Identity Formation" (1988) and "Discoursing the Others" (1991) are unpublished essays that were presented at the annual conferences of the *Journal of Curriculum Theorizing,* an interdisciplinary conference for teachers in schools and universities. In "The Animal: An Idea Waiting to Be Thought" (1992), also unpublished, he revisited developmental themes but placed more emphasis on the experience of animals throughout the life cycle of the human. In this essay he revisited the bear as a spiritual emissary and purveyor of transformation and renewal. In the short boxed piece, "Notes for a Diatribe on Masks," Paul succinctly explores the importance of animal masks in human culture.

In the next two essays and another short insert, we go back to Paul's past once more to illustrate the progression in his thinking. These pieces grew out of an independent study project during his undergraduate work at the University of Missouri with Dr. Rudolph Bennitt, his advisor and mentor. The two first met when Paul was a child on the Missouri State Experimental Farm, directed by his father, in Mountain Grove, where Paul and his family lived. On one occasion when Dr. Bennitt was visiting, Paul showed him his bird egg collection and was greatly impressed when the professor correctly identified eggs that Paul had mislabeled. After graduating from high school and completing his army service in the European theater of World War II, Paul entered the Wildlife Conservation Program directed by Professor Bennitt at the University of Missouri. Immediately Paul began what would become a three-year independent study project on animal eyes under Bennitt's tutelage. His assigned readings included Gordon Lynn Walls' *The Vertebrate Eye* (1942) and Vaughn Cornish's *Scenery and the Sense of Sight* (1935). After a lengthy exploration of the subject, Paul began writing a book on the eyes of animals. Bennitt's stern response—he repeatedly reprimanded Paul for making up words—may have caused Paul to abandon the idea of the book.

Out of that study, however, came a passionate and abiding faith in evolution and a poignant understanding of the importance of the physical changes in primates that accompanied unique primate perception. The essays dating from the early 1950s to the 1990s illustrate the growth of Paul's understanding as well as his literary voice. The first, "The Eyes Have It," published in *This is Nature, Thirty Years of the Best from Nature Magazine* (1959) and first published in *Nature Magazine* as "Eyes—Clues to Life Habits" (1951), is a traditional natural history essay on the relationship of animals' eyes to their habitat. "The Arboreal Eye" (1964) deals with the evolution of the primate eye and was written after Paul had begun his research on hunter/gatherers. This essay shows how the primate orientation to the world, as well as introspection, developed with bifocal vision and bipedality. The box entitled "What the Eye Knows," an unpublished piece written in 1995 toward the end of his life, is a playful distillation of his understanding of the evolution of intelligence and shows his penchant for synthesis.

After graduating from the University of Missouri and working for a year as field secretary for the Missouri Conservation Federation, directed by Charles Callison, Paul entered graduate studies at Yale University where he completed masters and doctoral degrees. In 1954 Paul accepted his first professorship at Knox College in Galesburg, Illinois, where, in addition to his teaching duties, he directed Green Oaks, the college biological field station. This was an extensive expanse of land consisting of old open pit mines as well as farmed land. With the help of students and faculty, he planted trees, transplanted indigenous aquatic flora and fauna into the ponds, and restored the fields to tallgrass prairie.

During those ten years at Knox College, he repeatedly rewrote *Man in the Landscape*. (The prolonged revision of the book, he insisted was one of the most frustrating experiences of his lifetime.) At this time, however, his interests did not remain fixed on revising the book but began turning toward our hunter/gatherer ancestors. His childhood experiences following two backwoods Ozark hunters introduced him early in life to the ambiguity among loving, hunting, killing, and eating animals and influenced his decision to study wildlife conservation at the University of Missouri. There the Cooperative Wildlife Unit, directed by Professor Bennitt, where his program was situated, included not only faculty but biologists from the Missouri Conservation Commission, the U.S. Fish and Wildlife Service, and the Wildlife Management Institute. In his introduction to *The Others* (1996), Paul explained:

> Everyone I knew there [in the Cooperative Wildlife Unit] loved animals, and yet they were all in the management business of killing them. Leopold wrote about land ethics and in the same book he spoke of . . . a shot duck dying in the morning sun. In the forty-five years since then, none of the people or programs for "saving" animals has seemed to me anywhere near as devoted and committed to the nonhuman world as those academics, mid-level bureaucrats, and their student candidates for jobs in state and federal game departments.

Out of this tension of "loving and killing animals" came his highly controversial *The Tender Carnivore and the Sacred Game* (1973).

The three essays that follow show his early explorations of this topic. "Reverence for Life at Lambaréné," published in *Landscape* (1958–1959), presents a harsh criticism of the stewardship role.

Although written in the 1950s, it is a primer for examining critically ethical arguments against killing animals as well as the rhetoric of stewardship. In this essay Paul also shows his propensity for questioning the premises of any author being hailed as an ecological visionary (as Albert Schweitzer was hailed after the publication of his book *Reverence for Life*).

The next two essays address hunting and the issues it raises. "A Theory of the Value of Hunting," presented at the Twenty-fourth North American Wildlife Conference and later published by the Wildlife Management Institute in Washington, D.C. (1959), remains to this day an important reference for hunting ethics. "Aggression and the Hunt," published in *Landscape* (1964), reflects his passion for the topic that was growing into the "hunting book." In the course of his research he had discovered *Meditations on Hunting* (1942), a philosophical treatise written by José Ortega y Gasset, the Spanish philosopher (1883–1955), a book that Paul considered invaluable. With his own book on hunting in its final stages of revision, he found a translator, Howard B. Wescott, for Ortega y Gasset's book and persuaded his publisher, Scribner's, to publish it. The first English translation of the book was published in 1972 with an introduction by Paul and is included here as "Meditations on Hunting."

As Paul had been carrying on an "intermittent meditation on the bear" for many years, it is appropriate that we begin and end this section on animals with essays on the bear. "The Significance of Bears" (1995) is one of his last unpublished essays and suggests the direction of yet another book had he lived on.

Paul's unique interpretation of human development was informed by years of research on the relationships of animals to human beings. "The human species," he observed, "emerged enacting, dreaming, and thinking animals and cannot be fully itself without them." This statement captures the understanding that emerged from his lifelong tracking of human/animal connections in all parts of the world.

FLORENCE R. SHEPARD

 # The Origin of Metaphor: The Animal Connection

For TWENTY YEARS I have had an intermittent meditation on the bear. I have come to see that the image of the bear represents far more than the animal itself, and it seems possible that the whole of the animal kingdom could be regarded as having a history parallel to the bear's biological evolution. That history is an elaboration of figures in the human imagination in which the animals became players in the emergence of human self-consciousness.

The noun "bear" comes from an Indo-European root term that gives us many other words, such as bury, borrow, burrow, bereave, bairn, birth, bier, which have to do with death and birth. As a process, the bear becomes a verb that linguist George Ruhl has called "one of the basic verbs of the language . . . irreducible" to further definition.

The dozens of meanings of the verb "bear" conform to one of three general meanings: to carry or transmit; to give birth; and to hold to a course—each with a place in cosmology. In the rich mythology of the bear, transmitting food and spiritual blessing takes place on earth, giving birth occurs in the underworld, and holding a course refers to the night sky.

If we suppose that these concepts and terms belong to a time when speech was in its infancy, the first glimmerings of analogy were made possible by using a limited vocabulary in multiple meanings, a time when abstractions demanded, as they still do, some reference to the tangible world. The bear was at the center of this transformation from natural history to cognitive history. The great bear was the best and richest gift of winter foods. She was the seeming virgin mother, bringing forth her young from the winter den, as though from the womb of the earth. And each bear was a traveler in 500 square miles whose timing so conjoined place and season that she seemed never to be lost and always in tune with the schedule and geography of all ripenings, hatchings, and spawnings.

Only from her natural history can we hope to grasp the cosmic scheme of Ursa Major, the constellation that dominates the northern sky, the passage maker around the Pole Star in Ursa Minor, a bearing taken by human travelers. Only from her natural appearance as the giver of life in bone, fat, glands, skin, and meat can we catch the significance of the sacramental meal. And only in her natural descent into the earth can we understand her guidance in the netherworld, from which birth and rebirth take place.

We ordinarily think of allusion to the cosmic bear in stories as illustrations of ideas. But I am suggesting the opposite: that these references to the heavenly bear, the giver of sacred food on earth, and the underworld genetrix are based on natural observations, made by our ancestors with scrupulous attention for a million years, as they began thinking about themselves philosophically. Conceptions of the spirit bear took shape from the actions of the animals themselves. Ideas about the structure of the word—about heaven and earth and the underworld—as performed by the bear became our way of grasping the human significance of those natural phenomena.

The great zoo of animal infinitives—to bear, to lark, to hound, to quail, to worm, to badger, to skunk—is likewise irreducible, because they are basic instruments connecting speech and consciousness. (By that I mean self-consciousness, because they are verbs that describe our actions.)

The conventional behavior by which language characterizes each species is of course, isolated, almost detached from the true animal. To quail, crouching tremulously before an oncoming danger, is but a single aspect of the life of a quail. How and when did we begin to abstract meaning in this way?

Perhaps it came from tracking, as though all quests converge on the horizon of forgotten time in some primal activity. As our ancestors became hunters, they plunged, late arrivals, into an old, savanna game of brain-making by means of clues. The evidence of seventy million years of mammalian predator/prey relationships in open country is given in the expansion of fossil crania, the stone signatures of bony braincases. The scenario is one of reciprocal, strategic pursuit and escape in which the amount of brain beyond that necessary for routine body functions is the measure of intelligence, which slowly increased in both predator and prey as they reckoned each other over the millennia.

As our forebears entered this long-standing counterplay among very intelligent carnivorous competitors and almost as bright, hoofed, prey species, let us imagine a sequence. Over time, mind refined the means of discovering and identifying the location of potential prey and dangerous competitors. At first the others were merely heard, seen, or smelled, their location enhanced by inference from the calls or actions of bystander animals such as birds. Then we were able to recognize droppings, nibbled stems, beds, and tracks, each with a temporal dimension: an age. In time we added the ability to discriminate individuals by sex, age, and physical condition from signs and to anticipate their direction, movement, and awareness of our presence. Add to this ambushing, running in relays, and cooperative stalking not only as skills but as conceptually sophisticated minding—as both predator and prey became sensitive to the daily and seasonal patterns of each other's movements and use of terrain—

players weaving themselves into an ecological and cognitive fabric. Finally, rehearsing, ritualizing, and planning, based on representations of the animals, mark the elegance of the human endeavor and bring us to the world of signs and symbols. The Others, at first recognized only in themselves, came at last to be hinted into existence, presences in their physical absence, to live as a body of signs.

But this is not all. Here is the crucial point: We brought to this world of insight and inference, to the natural and made representations, a distinctive primate preoccupation that might simply be called Self and Society. Whether in tranquility or a frantic interpersonal scramble, the higher primates are (and our ancestors probably were) ceaselessly appraising and interminably testing their status, membership, accessibility, and vulnerability within their own group. If the other higher primates had developed the figurative means of employing the images of other species in speech and art to represent their social concerns—they would be us.

So consider the hunting mind, surrounded by minute signals, tracking the Others by the signs that marked their identity, condition, activity, emotions, and health. Consider the hunter gathering bits and pieces that could be worn, danced, drawn, and abstracted to evoke ideas in a social context. In sum, consider a world of traces of animals living in the ecological community and providing an imagery that became embedded as the means for self-conscious primates to comprehend and articulate their own personal concerns. This idea of life as a quest among secret meanings and the perception of animals as a language about ourselves was not genderized in the conventional idiom of man the hunter and woman the gatherer. It was culture-deep and demanded prescience of all members of the human group.

It is not only in human evolution that the animals—and, in a slightly different way, plants—were essential to the emergence of mind but in the growth of the individual as well. The long course of our prehistory has shaped us, predisposed us to give attention in certain ways. There is the obsessive agenda of every small child, regardless of culture, in unconscious collaboration with parent or caregiver, to name the animals. The animal species system in nature is the least

ambiguous categorical model in the world. It is the doorway to cognition.

Category making in speech is the essential step to mind. Without
it no abstraction—indeed no thought as we experience it—is possible. This achievement in childhood is centered in the child's simultaneous interest in animals and anatomy—in eyes, ears, nose, bellies,
elbows—revealing an innate desire to perceptually dismember the
body (just as the hunter literally separates the parts of the body of the
prey), and to recognize unique characteristics of each part, so as to
distinguish one animal from another and one person from another.

As time passes in individual life, many of the transitions from one
state to another are, as it were, inconceivable, except as they are represented, embodied. The butterfly, the frog, the beetle, the egg, the
pupa, the birthing bear, the dying swan are such embodiments. Who
am I? I am he who, like the snake, sometimes sheds the skin of an
older self, he who now and then reemerges from an in-between state
of being briefly nobody. Like the cubs who come forth from the den
with their mother after a second winter, born yet again.

The thresholds between identities, those intervals of ambiguity,
are themselves marked by ambiguous animal figures: by the bat who
is winged but gives milk, by the owl who calls in the dusk between
day and night, by the fox who is confined neither to woods nor to
field. Gates, passages, entrances are traditionally guarded by statues of
these apotropaic animals, protectors and guides to transformation.
They are identified with the novitiates in the intervals of their non-
identity on the verge of initiation. They are the keepers of doorways
and thresholds. All such passages are temporary conditions of non-
identity or ambiguity, abstracted in speech and represented in ceremony by reference to the figures of species who live in the margins.
In the West we have reduced such figures to embellishments—the
lion and sphinx on the library steps, the gargoyles in the garden
pools—but in the wiser majority of human cultures and time these
guardians are seen as true spiritual powers and their natural forms are
respected accordingly. We still comprehend the spirituality of animals,
but we call it superstition. It is outside our rationality. In recent world

or otherworld religions those animals who live between realms are often demonized because of their ambiguous qualities, and we lose or neglect that significance for ourselves.

As we mature individually, our sense of self grows, expands, becomes dense. Under our skin we know ourselves as a dark landscape of desires and fears, peristaltic rhythms, the tumult of feelings to which we have given names but which have no forms. In meditative therapy, however, this invisible population becomes accessible— mediated by surprisingly autonomous animals who speak of the troubles of the heart, the gut, and the head as though it were natural that the centers of our inmost being were inhabited by animal guides.

In our adult lives, abstraction puts yet greater demands on the active imagery of embodiment. As verbs, the animals are like separate powers. But in order to flesh out complex ideas they must be combined. So it is that every society, every culture, creates composite animals for purposes of education and religious instruction. The sphinx, the angel, the mermaid, the minotaur reveal the indispensability of the animal figure in metaphysical concerns. Each society may deride the dragons created by other societies, as if they were illusory natural history, yet clings to its own as keys to the secrets of life.

At the three levels of our lives—self, society, and cosmos—there are the Others who save us from the deception of mirrors and the hopeless search for identity in our mere human reflection. Despite five centuries of humanistic insistence and a century of social science argument to the contrary, we are not our own creations. We do translate in creative ways from that vocabulary of animal reference— which first takes us outside ourselves so that we may then come back to who we are.

And why are they the best figures? Why not a language of machines, or a vocabulary of entirely "human" imagery, or simply abstractions without reference to any physical entities?

The answer is threefold. First, we are animals. We are distinct, yet share more with other animals than we differ from them. This overlapping, the difference in likeness, is normal and natural. Any definition that relies only on opposition, that denies ambiguity and the

Chinese boxes of plurality, can only, in the long view, be alienating and destructive.

Second, animals and plants are the middle ground between us and the nonliving world. They connect us to the planet and make us less lonely in the celestial universe. They mediate the inorganic aspects of ourselves. They are the common ground of ours and the earth's being.

Finally, animals vitalize all the important events and processes that make up our identity. They give life to our concepts and speech about ourselves, dispel the presumed superiority of mechanism, of bionics, of the enormous deception in electronic and mechanistic idealizing. So long as animals are the instrument of our cognition we will not surrender our organic connections.

In this world of escalating shortages and confrontations, a thread runs through the turmoil and crises. It is the definition of the "we" and "us" who feel ourselves at risk. The motor that drives the two dozen wars of ethnic conflict, the chronic alienation and criminality of youth, the extreme uncertainties of gender, and the environmental destruction is the question of identity. It is as though a plague of amnesia engulfs the world in which we suffer convulsions of desperate enactments of our possible selves arraigned against the Others. Intergroup conflict of one kind or another is tearing the world apart. Even our democratic process requiring a respected political opponent seems infected by this same epidemic of hatred of the Other. It is as though our war against nature were a kind of model—a kind of refinement of the fear of other species.

Insofar as self-recognition is an aspect of consciousness, cognition, and mind itself, my answer to the question of the meaning of nature is somewhat as follows: Just as the natural world provides us with the means of physical health—good air and water, nutrition, and healing substances—plants and animals are sensible figures in the health of the mind. Mind comes into existence as part of an evolutionary stream in which consciousness arises. Thought is an ecological activity, a process: We are recipients as well as actors in a world of Others.

In this process, the enigma of the self, or ourselves, is one half of a dyad. The other half is always an Other. Neither half of the dyad is comprehensible without that complementary half.

Our minds, like our bodies, still live in the Pleistocene. Nature is not scenery or the zoo in which the affluent part of the world seems to bask as though at the circus. It is the genesis of mind. The genesis of the mind, its dynamic, was a community of life that provided the cognitive terms out of which human identity arose, in which our sense of self continues to live.

We hear much these days about the loss of species and biological diversity, usually in terms of diminished ecosystems, destabilized environments, and the loss of unknown physical resources. I suspect that the greater loss is of another kind—the way a local fauna links the concept of the self and the uniqueness of place in different cultures. The loss of nonhuman diversity erases nuance in identity. We are coarsened by the loss of the animals. At the risk of being a little melodramatic, I close with a letter delivered to me by a bear:

> The Forest, the Sea,
> the Desert, the Prairie

Dear Primate P. Shepard and Interested Parties:
We nurtured the humans from a time before they were in the present form. When we first drew around them they were, like all animals, secure in a modest niche. Their evident peculiarities were clearly higher primate in their obsession, social status, and personal identity. In that respect they had grown smart, subtle, and devious, committed to a syndrome of tumultuous, aseasonal, erotic, hierarchic power. Like their nearest kin, they had elevated a certain kind of attention to a remarkable acuity which made them caring, protective, mean, and nasty in the peculiar combination of squinched facial feature and general pettiness of all monkeys.

In ancient savannas we slowly teased them out of their chauvinism. In our plumage we gave them aesthetics. In our courtships we tutored them in dance. In the gestures of antlered heads we showed them ceremony and the power of the mask. In our running hooves we revealed the secret of grain. As meat we courted them from within.

As foragers, their glance shifted a little from corms and rootlets, from the incessant bickering and scuffling of their inherited social introversion. They began looking at the horizon, where some of us were both danger and greater substance.

At first it was just a nudge—food stolen from the residue of lion kills, contended for with jackals and vultures, the search for hidden newborn gazelles, slow turtles, and eggs. We gradually became for them objects of thought, of remembering, telling, planning, and puzzling us out as the mystery of energy itself.

We tutored them from the outside. Dancing us, they began to see in us performances of their ideas and feelings. We became the concreteness of their own secret selves. We ate them and were eaten by them and so taught them the first metaphor of their frantic sociality: the outerness of themselves, and ourselves as their inwardness.

As a bequest of protein we broke the incessant round of herbivorous munching, giving them leisure. This made possible the lithe repose of apprentice predation and a new meaning for rumination, freeing them from the drudgery of browsing and the grip of relentless interpersonal strife. Bringing them into omnivorousness, we transformed them forever and they entered the game as a different player.

Not that they abandoned their appetite for greens and fruits, but enlarged it to seeds and meat, and to the risky landscapes of the mind. The savanna or tundra was essential to this tutorial, as a spaciousness open to infinite strategies of pursuit and escape, stretching the senses to their most distant reference. Their thought was invited to a new kind of executorship, incorporating remembrance and planning, to parallels between themselves and the Others and to words—our names—that enabled them to share images and ideas.

Having been committed in this way, first as food and then as the imagery of a great variety of events and processes, from signs in dreams to symbols in metaphysics, we have accompanied humans ever since. Having made them human, we continue to do so individually, and now serve more and more in therapeutic ways, holding their hands, so to speak, as they kill our wildness.

As slaves we stay close. As something to "pet" and to speak to, someone to be there and to need them, to be their first lesson in otherness, we have shared their homes for ten thousand years. They have made that tie a bond. From the private home we have gone out to the wounded and lonely, to those yearning for unqualified devotion—to hospitals, hospices, homes for the aged, wards of the sick, the enclaves of the handicapped and retarded. We now elicit speech from the autistic and trust from those in prison.

All that is well enough, but it involves only our minimal, domesticated selves, not our wild and perfect forms. It smells of dependency.

They still do not realize that they need us, thinking that we are simply one more comfort or curiosity. We have not regained the central place in their thought or meaning at the heart of their ecology and philosophy. Too often we are merely physical reality, mindless passion and brutality, or abstract tropes and symbols.

Sometimes we have to be underhanded. We slip into their dreams, we hide in the language, disguised in allusion, we mask our philosophical role in "nature aesthetics," we cavort to entertain. We wait in children's books, in pretty pictures, as burlesques in cartoons, as toys, designs in the very wallpaper, as rudimentary companions or pets.

We are marginalized, trivialized. We have sunk to being objects, commodities, possessions. We remain meat and hides, but only as a due and not as sacred gifts. They have forgotten how to learn the future from us, to follow our example, to heal themselves with our tissues and organs, forgotten that just watching our wild selves can be healing. Once we were the bridges, exemplars of change, mediators with the future and the unseen.

Their own numbers leave little room for us, and in this is their great misunderstanding. They are wrong about our departure, thinking it to be a part of their progress instead of their emptying. When we have gone they will not know who they are. Supposing themselves to be the purpose of it all, purpose will elude them. Their world will fade into an endless dusk with no whippoorwill to call the owl in the evening and no thrush to make a dawn.

The Others

 Animals and Identity Formation

Sometime before 100,000 years ago we began swallowing the animals. It became possible because, long before that, our ancestors drifted into parkland and savanna and extended our gut feelings for grass seeds and the Others as quarry and as danger and therefore as objects of thought.[1] The outcome had something to do with conscious reflections on them and us, or them as us, and on the deeper, nonreflecting black hole of our unconscious. The new minding and the animals became part of a mix that is with us still.

Carl Jung believed in an unconscious far more rich, versatile, and wise than the seamy pit of repressions conceived by Freud.[2] For the latter, those atavistic impulses broke from their dungeon into our awareness mostly as violent monsters. Jung agreed that monsters were indeed holed up in our psyches, but he insisted that a lot more went on there that could not be experienced as ravening beasts. He conceived of the unconscious as an animal mind, ours to be sure, in part that mind which existed before we broke from the shadows of the forest forever. Then the unconscious went on evolving in resonance with the peripatetic activities of the new, somewhat separate, "higher" brain and the new foraging itinerant way of life. The primate basement of the unconscious was arboreal and intensely social, and the hominids added to it the world of the omnivore—ingested, swallowed, in the two thousand centuries during which ecological strategy was the catalyst of the new, rational mind.

The existential problem, so to speak, was to keep the two minds in touch. We tend to think of this being done in dreams, myth, art, trance, and ritual. But James Hillman put it in terms, not of our action to this end, but of the means of the unconscious, the animal guides and messengers who mediate, nurture, heal, and generally conduct the traffic that keeps us from splitting.[3]

That the integrity of the self is somehow likened to a ghostly bestiary may seem unlikely. That they entered us on the model of ingestion and stayed on as a part of the minding of the world may only add to the implausibility. True, the evolutionary events I have suggested are conjectural. But the evidence for their present phenomenal reality is perhaps more credible. It relates to the ontogenetic process of becoming a self, constructed according to schedules, as appropriate experiences that satisfy a genetic predisposition.

Theorizing about body-image formation in infancy was given a full-dress discussion two decades ago by Wapner et al. in their symposium volume, *The Body Percept*.[4] They described the gradual resolution and differentiation of external body morphology by conscious speech and sight and touch as having a profound and unconscious extenuation. The conscious terminology of anatomical parts, along with the unconscious composite body map, together characterized an early stage of selfhood in which the infant names corresponding parts with those of its mother and simultaneously differentiates itself from her. One might wonder whether this dual activity does not establish the basis for a later metaphorical morphology of external world by the application of anatomical terms to the "body" of the earth. Such a landscape reference was extended to include the viscera by William Laughlin, an anthropologist, in describing the Aleutian use of the terminology of the organs of seals, walruses, and other large mammals when alluding to the terrain, as well as to a recognition of their own insides.[5] He noted that this language was nominative-functional, combining what in English would be subject and predicate.

The cognitive "butchering," necessary to define parts of an organism, undertaken initially when the hominids were invited to join the ongoing game of mind honing in the savannas, was termed "detotalizing" by Claude Lévi-Strauss.[6] The categorical or naming process of which animal systems are the model *sui generis* and the impeccable paradigm is inseparable from early speech acquisition and the calendar of ontogenesis.[7]

That the self and world inchoate are mutually and simultaneously constructed carries with it the corollary that inner and outer diversity have some sort of necessary dynamic, as Laughlin noted, going beyond the designation of static anatomical elements. As infants and children we hear much from inside, for example, all the lively sounds of the thuds and swishing of events. As the protagonist in Carlos Fuentes *Death of Artemo Cruz* wryly observed, the approach of death brought a new awareness of these sounds, ignored for seventy-one years, in their sudden silence.[8] And we feel these movements. Like mice in the walls, they are part of our yet-to-be-defined self implying creaturely life.

The cognitive apparatus of animal name learning, to which three-year-olds are resolutely addicted, includes the things animals do. I do not mean simply that they run and play, but more importantly that the nouns themselves become verbs in a multitude of infinitives: to duck, to badger, to outfox, to bear, to worm. . . . Perhaps these animate events are to our tissues what animals are to plants, and both to the bone what plants and animals are to the rock of the earth.

The animation of the personal interior rests on the reciprocity of speech and perceptual imagery (not always visual). The naming process, intimately prefiguring categorical thinking, is so intensely experienced in connection with animals that some additional dimension seems implied—that is, beyond the wonderful efficacy of the species system and the collective generic categories as the paragon of interrelated diversity. There is something more: a dimension related to shared skills and shared feelings, feelings that scamper across consciousness—a winged moment of joy, a scuttling shadow of fear, or a crawling anxiety. It remains for society, having spent a million years animal watching, to match this inner harlequinade with a natural world already held by myth to be a language—or for the individual, hungering inexplicably for the matching, to find the animals in the world that embody or enact each of many different feelings. There are many forms of this. Play, folktales, and dreams are three examples.

Consider play. Many organized, traditional games involve a mimesis: a pretending to be animals. These are games of predication in which conventional behavior becomes reference points in a linking of domains.[9] Among them are simple Leapfrog, Duck Waddle, Crab Walk, Itsy Bitsy Spider, Piggyback Riding, Hobbyhorse, Shark and Pom Pom in the water, Fox and Geese in snow, and all the other tag games evoking predator and prey. This is a psychodynamic strategy in which people take animal shape and, by becoming objects to themselves, are in a position to become subjects: a putting on and off of identity that makes for a sequence of matching of specific aspects of the self—movements and feelings—to images. Areas of the inchoate pronouns "me," "you," "they" are blocked out in the sense that actors rehearsing a play speak of "blocking" out their significant movements on stage. It is certainly unlikely that the total repertoire of animal games played by any group of children is sufficient to project/introject all the many ephemeral emotions that are part of the self: panic, envy, greed, cunning, tranquility, affection, shyness, and so on. It is the mode that is important. Like the desire to move one's hands over the surface of cubes, spheres, diamonds, and ovals as one identifies them, it is a practice gradually abandoned to the quickness of the eye, to that subtle stirring of the muscles which accompanies all speech in adults, and to the cultural elaboration of ritual motion in ceremonial occasions.

In folktales—in fairy tales especially—the dramatis personae are the inherent concerns of the child: anxiety about parental intent, small size, lack of skill; negative emotions such as envy, hostility, and aggression; sexual competition with a parent; fear of the dark; all are embedded and reified in stories with happy endings.[10] Their identity is often disguised in order to avoid the double bind and to give such intangible elements of the self a concrete sign image.

But what can be said specifically of the animals in fairy tales? They tend to be either bodily and instinctive aspects of the self or masters of transformation whose roles inspire acceptance of ourselves in a world ultimately good—as well as an unconscious apprehension of one's bad side, and the prospect of mastering it, a faith in the organic substrate and a rightness of being, relating even worrisome things inextricably with life's bright side. Like the Beast that Beauty must marry or the disgusting frog that the king forces the princess to wed and bed, the hairy, smelly, and slithery trepidations that at first appall her will become something helpful and beautiful. Or like the outcast son who had managed in his education only to "learn what the birds sing, the dogs bark, and the frogs croak," each in its own moment enabled him to become a wise pope. His faith in the things of the air, earth, and water as a meaningful pattern corresponding to his own nature, however inconsequential they seemed to others, would reward both himself and all Christendom in the end.

The role of animals in dreams, the third example, like that in fairy tales, has not been the subject of much study. Yet the evidence is suggestive. No doubt a dream animal can be itself—snakes may indeed refer to snakes—but there is substantial opinion that the figures are a code. As in fairy tales this may mean the disguise of social circumstances that are too close for comfort. They may also be animated elements of the self. Indeed, before the age of five there does not seem to be a separate self in children's dreams, but a sort of multiple perspective as though the self were dispersed among the congeries of images.[11]

Further, the frequency of animals in dreams diminishes from about sixty-seven percent in three-year-olds, to a typical seven percent in young adults—except in socially maladjusted individuals. The latter continue to have more animal dreams than happier or more competent schoolmates. Several interpretations are possible. One is that animals in dreams give emphatic and distinct form to a real but confusing variety of social processes and circumstances.

I resist the notion that we simply let animals "stand for" something and then grow out of it as we gain in ego strength. In cartoons, pretend play, and stories as well as in dreams, children are surrounded by a vast living sociality of creatures who, by the virtue of their ambiguity or paradox (like us, yet different), are agents of a modulating drama. Hillman speaks of dream animals as conductors to our own underworld. By this I think he intends to avoid the clinical diagnoses that would end in purely physiological processes and the misuse of metaphor as an intellectual (or, worse, aesthetic) trick. He keeps open the possibility of a reality of animal "power" that remains inexplicable, and he hints of polytheism.

Identity is not a simple resolution, for it has two aspects that seem to pull in opposite directions. In one direction is the savage portrayed by nineteenth-century classicists: creatures sunk so completely in synonymy that they are without personal selfhood. At the other extreme is the modern solitary soul: so elevated in the cultivation of individuality as to be anonymous, grieved in alienation and angst, isolated and without support. Only one type of concrete image is available that does not point in one direction or the other, simultaneously modeling both the plurality of uniqueness and the security of kinship. These animations help us to secure an otherwise refractory inner domain. They also serve as categorical paradigms with unlimited subtlety that undergird later, adult occasions for the use of animals as a referential adjunct in making the self whole.

Coming to terms with our inner person establishes an inner zoology accessible throughout life. Eligio Stephen Gallegos has pioneered a form of visual image therapy, for example, which calls for meditation on one of the seven energy centers along the spinal axis called chakras in Sanskrit.[12] The therapist guides the client in gently summoning the animal from each of these centers and inquiring into its well-being. In the patient's imagic state, a dialogue takes place and the condition of the animal is noted (deformed, wounded, caged, for example). Some may be poorly developed or larval, or underworld forms, or angry or frightened. As the therapy progresses, interaction may take place between the animals of different chakras, or they may change or swallow one another. Finally a council of all is envisaged. Throughout the therapy, which may take many sessions, the therapist does not interpret or explain but guides the subject in an interview— an inquiry in which the animals seem to speak for our respective centers of emotion, communication, intelligence, and so on. Watching a group of people with no prior experience in this therapy, simultaneously conversing with their heart chakra animal, one witnesses an astonishing demonstration of the individual's prior capacity for skills in gaining access to an inner ecology whose inhabitants may yearn to speak but who, unless embodied, have no tongue.

Although the foregoing examples seem to favor individualizing responses, that is not always the case. In the most vivid of ontogenetic moments—the initiation—there are superordinate incorporations. In the ceremonies proper, animals capable of transformation and renewal, such as the bear, are danced and the initiate is born into the social animal. In Lévi-Strauss's concept of totemic thought, Nature and Culture are conceived as a parallel series of complex interactions.[13] The cultural group is divided into a series of clans whose mutual obligations and functions are analogically related to a natural series of animal interrelationships among species for whom the clans are named. The analogy is a fiction that incorporates both myth and natural history as reference. The elegance of this system is implied in the lifelong vocation of biological systematics to which many totemic peoples are prone—that is, the elaboration of a taxonomy and natural history of hundreds or thousands of organisms as self-rewarding satisfaction. But the important point is that the clan system, like other fraternal and sororal subgroups in the name of animals, works toward saving the person from the isolating aspects of self-centered childhood individuation.

From these examples arise two questions having to do with how, psychologically, they work: What, if anything, does this have to do with the education of children? And what does this have to do with the real world of animals? Obviously, educational consequences begin long before school and, on the surface, are straightforward: There is the opportunity for a rich, diverse experience of animal images—coincident with speech acquisition and continuing in response to the child's own expressed desires—that establishes a living otherness as central (caring, knowing, responding, purposeful) lifelong metaphors. The insides of animals, and the eating of animals, by other animals should be a part of childhood experience as well—not because this represents brutal reality but because such experiences mediate an inside/outside dynamic and provide the trophic or ingestive model of relatedness that is the foundation for life on earth.

It is important that we reevaluate the place of anthropomorphism as an essential aspect of child perception. The disdain of modern biological science for "anthropomorphic error" in interpreting animal behavior may be justified in the training of scientists. But in the child, ascribing human characteristics to animals establishes a common ground or social illusion of ecological relationships necessary for our being "at home" on earth. Arriving at their formal education, children already have four years of the storied (or mythic) notion of the animals as people in their experience and, as well, a deep-set polymorphic zoology of the society of the self entrenched in the unconscious. If "nature study" means only the training of scientific protocol, then surely in forcing it on children we drive that sense of kinship—and the mysterious world of the self as a bestiary—into a limbo where it is increasingly inaccessible as an analogical resource in the lifelong pursuit of maturity as difference, diversity, and uniqueness framed in the confidence that, prior to it, there is a shared ground of living purpose.

Dreams, folktales, and play contain the patent error of pretending that each animal corresponds to a single conventional attribute: slyness in the fox, voraciousness in the wolf, lumbering clumsiness in the ox, goodness in the dove. True, this single dimension does not do justice to the intricate lives of these animals. But the years from one to four are committed to the study of real animals—rather, to something more important that creates an inner resonance with similar experiences later in life. Indeed, the child from five to twelve can profit from nature study and is perfectly capable of keeping alive the dual reality of story and nature.

What is the role of real animals in all this? On planet earth, we are on the threshold of cataclysmic destruction of the world's animal life on the order of extinction exceeding that of the Pleistocene or the close of the Cretaceous geological periods. As a naturalist I sometimes worry that I am attending to sign images: iconography important to the mind but independent of those creatures we can cook and eat or be eaten by.

While I suspect there may be some profound consequences for encounters between small children and flesh-and-blood animals, I cannot say how such encounters affect the development of what Gallegos calls the "personal totem pole." Perhaps they give it extraordinary intensity. But turning the question around, I strongly suspect that the childhood development of an animal sign imagery has effects on our adult perception of animals and our decisions about their fate. In other words, human ecology is deeply influenced by the animals of our imagination. Our early experience of animals is the source of the consanguinity that Cassirer[14] observes is the basis of all mythology—that is, our sense of kinship to the Others. That kinship is not simply an ecology or a moral abstraction. It is a family tie and carries responsibility. Neither a philosophical ethic of the "rights of animals" nor a scientific thesis on the stability of natural systems in the long run will justify the enormous cost of saving Nature without a prior sense that the Others are somehow us.

The human personality is constructed by the discovery of its own elements in the external world—an epigenetic process. Further, I have speculated that this ontogeny is a result of our species, natural history, and evolution. In this animals play a unique role. The differentiation of our selves from most things in our environment has a terrible, exacting, *distancing* about it. But the animals, having family resemblance, are both like us and unlike us; they are ambiguous. The acceptance or rejection of this ambiguity, and of all ambiguity, offers the greatest possibilities for becoming more fully human.

NOTES

1. Harry Jerrison, *Evolution of the Brain and Intelligence* (New York: Academic Press, 1973).

2. John Freeman, "Introduction," in *Man and His Symbols,* ed. C.G. Jung (New York: Harper, 1979).

3. James Hillman, "Animals," *The Dream of the Underworld* (New York: Harper, 1979).

4. Herman A. Witkin, "Development of the Body Concept and Psychological Differentiation," in *The Body Percept,* ed. S. Wagner et al. (New York: Random House, 1965).

5. William Laughlin, "Acquisition of Anatomical Knowledge by Ancient Man," in *Social Life of Early Man,* ed. S.L. Washburn (Chicago: Aldine, 1961).

6. Claude Lévi-Strauss, "Categories, Elements, Species, Numbers," *The Savage Mind.* (Chicago: University of Chicago Press, 1966).

7. Eric Lenneberg. *The Biological Foundations of Language* (New York: Wiley, 1967).

8. Carlos Fuentes, *The Death of Artemo Cruz* (New York: Farrar, Strauss & Giroux, 1966), p. 83.

9. James Fernandez, "Persuasions and Performances: the Beast in Every Body and the Metaphors of Everyman," *Daedalus* 101(1) (1972):39–60.

10. Bruno Bettelheim, *The Uses of Enchantment* (New York: Knopf, 1976).

11. David Foulkes, *Children's Dreams* (New York: Wiley, 1982).

12. Eligio Stephen Gallegos, *The Personal Totem Pole* (Santa Fe: Moon Bear Press, 1988).

13. Joseph Henderson, *Thresholds of Initiation* (Middleton: Wesleyan University Press, 1967).

14. Ernst Cassirer, *An Essay on Man* (New Haven: Yale University Press, 1944).

Discoursing the Others

O F ALL THE HUMAN CHARACTERISTICS that have risen and fallen, over the centuries, intended to differentiate the human from other species—sentience, thought, speech, learning, spirit, feeling, culture, intelligence, memory—the only one surviving the onslaught of the sciences is the highly specialized form of reflexive consciousness in which we constantly assess our "place." Pronouns are at the center of our reality, our identity, and our relationships to others. Strangely, though, the pronouns are not self-evident but must be perceived in anomalous, coded forms, distanced as the Others, before being internalized. This prelude of projection is essential to assimilation. This provisional and subjective separation from the Others, a disconnection from absorption and suffusion, is what our species is about: a dangerous adventure in radical disjunction. Recovery from this objectivity—a biomorphic affiliation—is the lien on sanity, the necessary response.

In "space age" cruising with its alter ego, the whiz-bang submolecular adventuring, we reach with great hope to the inanimate cosmos. But the "wonders" of the celestial universe only divert us from the intolerable, inchoate self. At the other end we observe populations of brief subatomic entities. Both preoccupations mask a lonely world where our inability to overcome the numbing terror of perfect estrangement is submerged in the entertainment of space travel imagery, with its dreams of "alien" others, or in the ritual pronouncements of the priests of anesthetizing physical abstractions. Their veracity is not the question: Neither their mega-verse nor the micro-verse offers the slightest clue as to who we are.

The problem is that the enigma of the we-I-you-they came into existence, not in the phenomenological world of quarks, quanta, or galaxies, but in reference to other beings. The otherness of gluons, like the otherness of suns, has no homology to our experience. The otherness of our fellow humans is not in their natures but in their roles, and therefore their identities are as elusive as the differing states through which we individually move in a lifetime.

Between us and the rock and water and air of this planet, which are the nearby epiphenomenal expressions of that cosmos of the mullahs of physics and astronomy, between us and that impenetrable mystery, are the only others of our true grounding. It is their diversity which is the context of that final, incalculable quality: the twofold state of likeness in difference. It is not that people and animals are analogous—although that is a source of the myths from which structure is created—nor that we are embedded in them as a species in the concentric circles of genetic kinship—although that is the organic basis of our intuition. It is the nature of binary reality—as , for example, in the almost unimaginable moment of migratory flight conceived against the familiar joy of family compassion, one so surely common as to epitomize the emotion in ourselves, the other, strange beyond our imagination. In what is shared we are linked to that which is not shared, and thus to the horrendous stolidity of the stony island on which we live. It is the simultaneous inwardness and outwardness of the animals that gives us the heart to be.

As for their own identity, the animals all seem in turn to have no doubt. The fissure that we have made our own, dividing us from them, does not vex them. A naturalist may see occasions as I did when a blue grosbeak and an indigo bunting hesitated on the same branch, each reflected in the other's blue mirror, an instant when the other and the self seemed uncertain in the private sense of respective identities—but only for a second. (Their equivocation was itself endearing.) The assurance of the Others, not their lapses, gives us substance. They are the talismans of authenticity, affirmations of categorical certainty.

Our own enigmatic identity is the captive of culture—that intraspecies capacity for invented and hypostasized membership and exclusion. This callow state is the mark of our neoteny, the necessary indetermination upon which our psychology is based and the experiment in primate-human organization on which our species has created arbitrary categories of admission and omission. We identify ourselves by political and professional affiliation and opposition, by communal and family ties, by the making of an evanescent web of preferences, a tissue so frail that we must constantly attend to its repair. It is possible—this cultural frame of formal inspired pronominal identity—only because the other species are unequivocal. As intermediaries they are therefore the mediators, the presences on one side of a discourse.

Educational experience is constructed through multiple discourses; among ourselves the dialogue is speech. But the discourse with the Others—in which we find the avenues of the first and last education: the recognition of our being as this living but nonhuman otherness—is not speech. Those who toil to teach chimpanzees elocution and dolphins semantics, or to translate their imagined soliloquies, pronouncements about life that would reveal them to us and thereby us to ourselves, may have misconstrued the discourse between species.

It is intuitively obvious that we exist in a continuum with the earth. But the phenomenal evidence is not apparent except as a series of living Others, the elemental forms of which demonstrate their continuity irrefutably, while the close links signal the primate steps by which we have become objects to ourselves. Within the framework of earthlings, the Others further model our diversity to us. Only the categorical elegance of the species system can do this as the final social metaphor. Their multiplicity is essential to our consciousness. The multitudes—of ourselves and the Others—and their diversity counter the radical disparity of incomprehensible absolutes of time and space, the insanity of utter rational analysis into electrons and black holes, or the infantilisms of undifferentiated bliss.

Like all discourses, this one is ultimately energetic. Yet the cyclic flow of materials is all abstraction to us. What we experience in the presence of animals is the consummate tableau of reception and reciprocity. Having understood that they are analogous, we comprehend other forms of symbiosis that are at once the source of vital activity and the fountainhead of meaning. A mind so shaped as ours is at liberty only to construct the narratives which satisfy the requirement that the outer world is the stuff of which the great "I am" is made. It may do so from the vapors of its own caprice, from the nightmares of black holes and ultimate particles, or it may do so in the context in which the mind came into existence: in a complex community of living Others.

The Animal:
An Idea
Waiting to
Be Thought

MOST PEOPLE, during most of human existence, have reflected on the animals they eat and, then, in their responses, reflect something of the animal in themselves. When we eat animals, we take not only the body but also some of the being of the animal into ourselves. We surround it, watchful. Thoughts about food are really about assimilation and correspondence, messages between inside and outside. Eventually, even thought about final things comes back in some way to questions of likeness and difference.

Of the many rhythms of life, the passages between inside and outside have multiple expressions. Like caravans in transit across the landscape, the elemental crossings are those of physical substance. For the alimentary and respiratory systems are the royal road connecting the two great unknowns in every young life, self and world. The sensory systems have their own trails and runners without cargo, bearing only messages, correspondence.

In later life, the inner and outer come to be understood as a whole and portions of the terrain, the fauna and flora, take on accord with an inner prospect, whole systems of meaning, abstractions nurtured throughout by the tangible reality of food. But we begin as children taking in animals by the most intense and compulsive scrutiny. Watching and naming are a kind of nutrition. In the eyes of children each animal is unambiguous, as plain and literal as its name: horse, cow, dog, chicken, bird, elephant. In being itself, each animal stands only for what it is and moves according to its kind. Each is part of a fauna of conventions: whinnying eagerness, bovine nurturing, yapping pursuit, clucking anxiety, and aerial capering. In its characteristic act, each is a condensed feeling, fear, idea, or emotion. Even though these may be unconscious or unformulated experiences, they may be unassimilated in the human watcher and listener. The animals—told in a tale, enacted, and mimicked—are references whose twins inside the person can be, one by one, brought to life. Attention to this predilection has all the power of nourishment itself, born of curiosity, naming, and imaginative play.

All games begin in that enactment. Beast by beast, in the first years of life, emotions, feelings, attitudes, and intentions take their place in the fauna of the self. Freed of explanation, they remain ambiguous. Finding oneself in other humans will not do as a training ground of perception. Our quicksilver shifts of mood, shading of traits, conflict and blending of responses, tempering and concealment of motives, fluid shifting of inner shapes to fit circumstances, are too slippery. Human subtlety is important later; but as a training ground of perceptions it is madness.

The decade from the beginnings of speech to the onset of puberty is all we have to load the ark. Its zoology must be unequivocal and without recondite allusions. Poetry, song, and game must mean what they say and must be clearly played, as conspicuous as a cat pursuing a ball. It is right to mimic the fox and goose in the tag game of feigned capture or to speak the lines of the Little Pig and Chicken Little. In doing so we become connected to common ground with other life in spite of external differences between us. Anthropomorphism is essential: the true but subtle means of inter-species communication, full of invisible nuance, as removed from sensory detection as the genome itself. In pretending that animals speak to one another, the child imposes on animals a pseudo-humanity which, although illusory, is the glue of real kinship. In such farces of socialized ecology the vital natures of animals are encountered and become our defense against the conspiracy that animals are only machines or artifacts—and, therefore, our defense against the lie that we ourselves are made of cogs, wheels, and wires. It is also important to see the insides of animals, for organs have names too, forming a fauna of stomachs and lungs and hearts, to which ours belong.

Much is at stake in the first decade of life, for it culminates in bonding to the earth matrix, which is sandwiched between infant/mother bonds and formal entry into adult membership in a cosmos. Here the foundation for a poetry of ultimate meaning is based. In this matrix or mesocosm, animals are the animate aspect, the speakers. Play forms the bonding matrix that joins natural history, observation, and speech. The richness of one's eventual personal philosophy depends on it. But during childhood abstractions of chemistry, physics, ecosystems, morals, and ethics are poisonous to the imagination.

Animals, with their capacity to bear messages and move children, do not live in empty space—except in the zoo or the barnyard. Like their bodies, their location is unique. Habitat is the literal space of the ground of thought. As messages, animals come into thought trailing the dust of that association.

The home range of ten-year-olds is the first context of spatial and temporal thought—perceived unconsciously as a replica of their mother's body, the terrain of the newborn, as it moved like a traveler locating the spatially ordered reality of life. The living terrain is conceptualized by the child as an ordered space inhabited by creatures and their signs—turtles, frogs, bird nests, mice, rabbits, tracks—an event-world in which animals are the players and the pattern of their compelling actions are like moments in the life of a great spatial being.

The end of childhood is the end of that simple identity. The literal fauna become the external expression of the child's own congeries of feelings and bodily processes: a community of self-confidence. That confidence will be tested throughout the lifetime of the individual, for life is full of contradictions. Indeed, adolescence is a preparation for ambiguity, a realm of penumbral shadows. Its language includes a widening sensitivity to pun and poetry. Appropriate to its psychology is attention to the zones between categories. These margins have animals, too. The borders from which obscenity and taboo arise are figured in the creatures who give embodiment to overlapping reality: the insects who crawl between two surfaces, the owl flying at dusk, the finless marine life, the legless land animals, the bat who seems to be both bird and mammal, the pangolin, the fox, the hare. . . .

Adolescents are themselves marginal beings. Between stages of life, they are on the shifting sands of an uncertain identity. In this respect their symbols are the changeling species: the self-renewing, skin-shedding snake; the amphibious frog who loses a tail and grows legs; the caterpillar who is metamorphosed into a butterfly. In each the thought of a new birth is manifest, the concrete expression of transformation. Human psychogenesis is such that the adolescent is, for a time, plugged back into his own natality. The concreteness of life, literal in the maternal and natural matrices, given consciousness in speech itself, will be re-viewed in a new metaphorical idiom.

No echo of this infantile state is more crucial than nourishment. Eating—the most fundamental route from outer to inner—is to be reevoked as the ritual act at the core of transformation and relatedness. Its emotion is refocused in intellectual and symbolic ways using incorporation as the metaphor of connectedness. Henceforth all rituals of transitional states—the reception of spiritual life, elevation in social status, marriage—are celebrated or represented as feasts. Sacred meals, taboo foods, and dietary laws everywhere refer to how what is eaten is an agent of change in the eater.

In the transformation from child to adult, animal messages signify

new birth. But they perform more broadly than puberty and ceremonial initiation. As a collective, animals of the natural environment
together comprise the concrete metaphor of the human group. In
tribal culture, each clan is committed to a particular species. This
species, because of its ecological relationships to other species, provides a vehicle for the dynamic of myth and the rules of society.
Together the clans constitute the whole in a way that is analogous
both to the ecology of animals and to the synthesis of the self from
animal behavior observed in childhood. In myth the ecology of
species is seen as a fictitious society, a kind of holding ground. The
same fauna mimicked in childhood play is liberated in maturity into
new levels of social and metaphysical deliberation. An example is the
construction of dragon beasts that are spliced together from different
species to make images for complex ideas.

The use of animals in play in the first decade of life gives way in
the young adult to dance—a universal human activity derived from
the rhythmic imitation of animals. In traditional societies, a particular human group acquires a style of dance of its own, uniting its
members while, at the same time, affirming the tutorial role of birds
and mammals. Humans have always suspected that certain animals are
the masters and keepers of important secrets: metamorphosis, birth,
puberty, healing, protection, fertility, and food getting. By dancing the
animal, we assimilate these mysteries into adult understanding and
recover them as a power of humankind.

Part of becoming adult is the dawning realization that the principle of transformation is a major feature of the cosmos. Movement
and passage making are inseparable from time consciousness. In our
modern adult life—other than in hunting participated in primarily
by men—there is a dearth of experiences with wild animals. In tribal
cultures, dancing in the feathers of birds and wearing the masks of
mammals displays the shape-shifting capacities of the soul. The religious principle of altered states, like that of boundary habitation, has
its special animals, the greatest of which, in the northern hemisphere,
is the bear. From its natural history comes a rainbow of homological
suggestions so powerful that it may have changed the history and
evolution of human thought.

Many features of the bear—especially the many races of the brown bear—place it in correspondence to humanity. Its size, appearance, mobility, dexterity, omnivorousness, reproduction, annual cycle, length of life, social behavior, and intelligence have an eerie relationship to our own. These characteristics are the source of enduring speculative analogy and psychological tension. The geography of this rapture is as wide as the distribution of the brown bears and as ancient as mankind: a whole paragraph in the zoological hieroglyphics of human consciousness.

The bear is the only familiar omnivore whose size approximates our own. Omnivory is not only a kind of diet but a versatile style of perception—exploratory, pushy, relentless searching, analytic, and risk taking. It tastes the fruits of all action, the meat of all situations, the kernel of all experiences, the root of being. The bear is fisher, hunter, berry-picker, bulb-digger, honey-gatherer. It has an expressive face, binocular vision, vocal and gestural responses, sitting and bipedal stances, almost no tail, and fine dexterity. Mother bears give birth secretly, tend and teach their young, and defend them fearlessly. And yet the bear is vividly other: huge, furry, long-muzzled and clawed, quadrupedal, in these things nothing like a human.

In winter the brown bear withdraws into the earth from which it came as a cub. This winter sleep coincides with the death of nature. Spring comes after the bear's emergence, as though called back to life. This passage into and out of the earth becomes a rhythmic, seasonal movement. Winter sleep of bears is a repetitive pattern, like courtship and migratory patterns in birds, that urges us to watch and understand them and invites us to mime and dance their essence. The bear's journey into the earth, translated into the rhythm of the life cycle, is unmistakably about death and reincarnation.

For perhaps fifty millennia, festivals of the bear ceremony have followed this premise: The bear is a sacred messenger and mediator, purveyor of meat, paradigmatic grandparent, teacher, traveler between worlds. In part the message of the hunted bear is this: Save the bones of your dead and inter them in the earth. Remember that the spirit survives and lives again. Connect this sacred quality with every individual in a ceremonial feast of communion. There are parallel lives below and above your plane that are eternal. Passage between them is the ultimate movement by which you know life.

The bear is the keeper of all gates—between life and death, this world and the others, flesh and spirit, human and animal form, inside and outside, even the phases of individual life—as well as the mediator between man and woman, the natural and the sacred. Leaving the animal out of the myth of resurrection is a historical development, the collapse of an instructive metaphor. Replacing the bear exclusively with the human figure denies us each our bearskin. Zealously repudiating the animal form, omitting the middle matrix, it retreats from the polymorphic ambiguity of life. Juvenile literalness of a bearless cosmos deprives us of personal experience of the sacred paradigm, substituting for it an abstract, verbal exegesis. The loss makes for autism, middlemen, desperation, the failure of the kindred species who think in us.

Carrying a positivistic, literal attitude toward animals into the adult sphere marks the failure of initiation and maturity in human life. The totally humanized myth of immortality is part of the zeitgeist of domestication, its ritual centered on sacrifice rather than the sacred hunt. Our dreams, however, remain true to a different world. Hunger for the wild animal's significance is reflected palely in the vicarious imagery of toys, decorative arts, virtuoso and eccentric originality, pets, and media stereotypes. No fine words can replace the dances and feasts of participation. These acts remind us that we were thought up by the different beasts.

They are kindred and ancestors. Before men existed they worked out the round of life in thousands of variations as though anticipating the needs of style in the experiment of human cultures. Like the bear we are selves composed of sleeping figures, each a secret that can be awakened in acts of correspondence. Self-consciousness is possible only in a world of Others. We are members of a human family and society, but the presence of animal Others enlarges our perception of the self beyond our immediate habitations: not only to the limits of the outer world but deeply inward to that ground of being where live the lizard, the monkey, and the fish.

Notes for a Diatribe on Masks

+ Unlike many other animals, primates are preoccupied with faces. In faces the whole world of feelings, intentions, actions, and powers is revealed.
+ The powers are those that animate the cosmos.
+ The text for learning and reading the play of these powers is given concrete form in the species of animals.
+ Each animal enacts a power and is represented by a masklike face.
+ Individual humans have all of these powers or species within.
+ But we can best see the inner play of these powers in the interplay of the animals. They are *our story.*
+ We perform segments of this story (wearing masks). The dramatized myth is also the enacted metaphor of ritual.
+ The mask transforms the actor into the power. In it he/she becomes. The mask is the concrete sign of the paradox of all transformation.
+ Its meaning is exhausted in its features—it does not disguise, except in cultures that reject malleable identity.
+ Dispersed powers, sacred significance in animals, mutability all are counter to monotheism.
+ Monotheism is an ideology that regards masks and animals as diabolical or toys, belief in appearance as "gullibility," separating essence and idea from reality.
+ And play is never deep play.
+ Adult play is understood as idleness or recuperation/recreation from work and duty. There are no flutes, dances, plays, or playfulness in Greek philosophy and monotheism.
+ Make-believe play and improvised identity in children are tendencies inherited from archaic societies, learning to suspend disbelief, enabling us to understand that a person wearing the face of a frog is the frog power and spirit.

The Eyes
Have It

IF SOMEONE WERE to put into your hands an animal that you had never seen before and asked you to describe its behavior in the wild, what about it would provide you with ideas?

One of the best organs to examine for clues to the animal's activity would be the eyes. Written in them is the story of the creature's life. They are also the diary of its racial history: the habits, food sources, and fears born on the ancestral experience of its kind. The eye is the soul's window, yes. But it is cast in a mold fashioned by particular habits from universal material.

Among vertebrates, the fishes developed the first good eyes, by our standards. These underwater eyes were nearly all alike in many respects. When a land environment was adopted in the process of evolution, however, eyes became as divergent as the demands upon them.

Some eyes were sharpened to perfection by natural selection, only to become debased and feeble through the vicissitudes of geological time. The fish eye reached an evolutionary end because sight under water is limited. Snakes took their fine reptilian eyes beneath the ground and nocturnal mammals wasted them in the darkness of night. When the descendants of these animals once again moved into the daylight they were handicapped, and the eye was remolded.

What were the forces of this mold? What has it meant to the eye that some creatures must escape in order to live, while others must capture? Why does the rat see only shadows, while the hawk watches a grasshopper a half-mile away? Why do some animals see color and others not?

In the beginning, ocular raw materials were much the same among all vertebrates. But life has since demanded many modifications, and these in comparatively short periods of geological time. If we assume that a "better" eye is one that resolves an image into greater detail, then a strange classification has come into existence. The development of good sight is not in accord with our system of systematics. "Higher" and "lower" groups, each, have all sorts of eyes. The magnificent eyes of birds are belittled by the inadequate ones of the kiwi, and the blurry vision of mammals is honored in the good eyes of humans. The reptiles, amphibians, and birds all harbor their paragons of vision and the opposite.

Among various kinds of animals, visual efficiency is in proportion to the importance of sight to the animal's existence. A golden eagle hunting a newborn relies on acute vision. The antelope, too, is a creature of much daylight activity and is often dependent on keen sight. But the antelope may also *smell* the coyote that stalks it. The coyote may put eyes, nose, and a superior intelligence to work in stalking an antelope, stealing poultry, eating sheep and carrion, or digging up rodents such as the pocket gopher. To elude the coyote, the gopher depends only on tactile and hearing senses. To the gopher, sight has little or no survival value. This animal is at the bottom of the visual scale, the eagle at the top.

Animals at the top of the visual ladder—predator or prey, fish or furbearer—have large eyes; not large in proportion to their heads or bodies, but in absolute size. This is because the seeing cells of the eye, which receive an image on the retina, are all about the same size, whether in bobcat or bison. A big eye is important, then; it holds more such cells, which break down the picture into greater detail. The big eye also increases the distance between the lens and retina, throwing a bigger image on the living screen.

Bright light enables an object to be seen better and at greater distance. Thus the best eyes belong to animals of daytime and *must* have daylight to function well. To most temperate-zone reptiles, sunlight is the great motivator. What use are eyes that see in the "dark" when nights are spent in immobility? Many mammals, and some birds and amphibians, have developed an eye that works fairly well either in daylight or night. Some creatures' eyes shine in the dark, but the eye shine is nothing more than a reflective material that helps make the most of whatever light strikes the retina. In daylight some of these animals cover this material in a manner similar to closing venetian blinds.

Generally, nocturnal animals depend on senses other than eyesight; they have poor eyes. Some find it necessary to see, like the owls, and develop exceedingly sensitive eyes. Most mammals are nocturnal, probably because their ancestors were hunted down in the daylight. It is mostly warm-blooded creatures that seek the cover of darkness for protection, like the mice who efficiently scavenge our kitchens. Early mammals inherited keen vision from their reptilian ancestry— good eyes like those of today's lizards. But in millions of years of sneaky existence in the dark the eyes degenerated, as is shown by any mole or shrew.

Predators generally have greater stamina, quicker reflexes, or move more rapidly than their prey. High mobility requires good ocular resolving power and rapid focusing to avoid collisions. The size of an animal's eyes is directly proportional to its speed. Without bigger eyes the bass and trout, for example, might be relegated to the food habits of carp and catfish.

Fast-moving animals also need quick adaptation to changes in light intensities. A powerfully muscled iris (which stops down the pupil) provides extensive dilation and contraction for just such reasons. Cats see well in the "dark," but also enjoy sunbathing. This luxury would be impossible without an iris muscle that contracts to a tiny slit and protects the sensitive retina in strong light. A Cooper's hawk, darting from the dark woods into a sunlit clearing where sparrows feed, or out of the midday sky to capture a flicker on the forest floor, flies through drastic changes of light intensity. The same is true of a sea lion diving for squid or fish. Hawk and sea lion are equipped with a flexible iris.

Accurate and immediate estimation of distance is also important to a predator. The striking snake, leaping tiger, or diving falcon must judge quickly. Binocularity—or placement of the eyes in the head so they can be focused on a single object simultaneously—makes such judgment possible. This placement—*frontality*—is characteristic of the predator anatomy. Prey species, on the other hand, have lateral eyes, each with its own field of vision. Many occasional predators, who both hunt and are hunted, fall between the two extremes. Their oblique eyes combine the values of binocularity with the advantage of a wide visual field.

Every animal falls into this picture somewhere. The cottontail rabbit, a prey species, does not see the grass it eats. *Binocularly* it sees only a few degrees in front and rear—where it may need distance estimation to dodge an immediate pursuer. But with one eye on each side, the rabbit has a total lateral visual field of 360 degrees. Nor can an enemy approach unseen from above, as the rabbit also sees up without moving its head. Opossum eyes are a compromise. The opossum may capture a small invertebrate one moment and scurry up a tree for its life the next. The cottontail's eyes are 90 degrees off the body axis; the opossum's are 30 degrees off. The lynx, with few enemies to fear and stalking proficiency imperative, is totally frontal, with eyes in the body line.

Eyes also become dorsal or ventral. The woodcock, probing the mud for larvae, watches the surrounding undergrowth with large eyes far back on the head. Although a predator as it feeds, the woodcock has little need of seeing the mud its long bill explores. The turtles, eminently successful through the ages, have little to fear from above. Their eyes are almost frontal and canted downward to scan the bottom for food. It should be noted that frontality may have other values than for predation. The eyes of primates are frontal presumably because of close-up use of the hands and the tree-swinging habit.

The perception of motion is a complex part of seeing, but one aspect is of special interest regarding food habits. It has been asked how a hawk, sitting in a dead tree, can spot a mouse moving anywhere in the field—unless it happens to be looking right toward it. The registration of movement on the retina is exaggerated around the boundaries of the visual field, as though to compensate for not "looking that way." This seems to be especially true of the lateral periphery. You can demonstrate this for yourself when you are a passenger in an automobile by tipping your head horizontally and watching the road ahead; it seems to rush beneath the car at increased speed.

The value of color vision and its corollary, protective coloration, is a popular topic for debate. Color vision is scattered about the tree of life without certain connection to predation. But the only groups that use sight *exclusively* for food-getting seem to possess it throughout: the birds and lizards, for instance. It is also found in the turtles, the higher fishes, and the primates. The contrast value of color is obvious. (What if all your books were one color?) It would seem valuable to a predator that must discern its prey. Yet protective coloration appears to compensate for the increased vulnerability of the hunted animal—which, in many cases, does not see color itself. If such coloration were completely successful it would appear to make color vision in the predator a liability—witness the success of "color-blind" artillery observers who spotted camouflaged installations from the air during World War II. Probably the two forces, like the "balance of nature," are never quite in adjustment.

Predation and habitat have interworked to produce combinations of eye characteristics. Most of the organ's evolution was completed by the time vertebrates came on land. The terrestrial environment superposed modifications on an organ fashioned under water in an extremely different set of conditions. The higher fishes have superlative eyes, but the development of aquatic vision is limited by the medium itself. Visibility has a range of about 150 feet in the clearest water, and no change in the eye can alter this. In fish, the lens does all the focusing and the cornea serves only as a protective window. The cornea's refractive index is the same as that of water, so that light rays pass directly through it. But not so in the air. Rays entering the eye bend once at the cornea and again at the lens, enabling the terrestrial eye to develop a small, efficient lens. The bulky, spherical lens of the fish bumps against the cornea as far forward as possible in order to attain a wide visual angle. It protrudes through the pupil blocking its contraction. This, as we have seen, puts a limit on light tolerance. Only a few fishes, such as sharks, have pupil mobility of the sort that makes cats and seals independent of day and night.

Many animals that feed or live in the water are faced with the necessity of seeing well in air, too. Without compensation, their air eyes are sadly farsighted under water, for the cornea is no longer refractive. The problem is met in four ways.

The first method is the nictitating membrane. Aquatic birds utilize the nictitating membrane when diving. It is a thin, third eyelid, underlying the regular eyelids, and found in all birds. In these diving ducks the membrane has a clear window of highly refractive material. A canvasback, diving for submerged plants, closes the third eyelid to bring the food into focus. The loons and auks share this device with the ducks. It is supplemented by a powerful iris muscle.

In many creatures a strong iris muscle constitutes the sole means of focusing under water. This is the second method. This iris muscle is attached to the lens so that, when it contracts, the lens is squeezed out of shape and into focus. The turtles, otters, and cormorants have these soft lenses in the grip of a vigorous iris muscle. This lens-squeezing is typical of all reptiles (except snakes) and birds, although greatly accentuated in the divers. Early in their history the mammals lost this efficient method of accommodation, or focusing. It was one of the penalties for night-prowling. Their iris muscle gradually withdrew from the lens, degenerated, and was replaced by zonular fibers, which guide but do not squeeze the lens. Under this handicap, our limited focusing is based on the elasticity of the lens and contraction of the zonular fibers.

The kingfisher, and at least one fish, show us a third way for air eyes to see under water. The kingfisher possesses two foveal areas: regions of keen sight where the image is focused on the retina. They are different distances from the lens. The more nasal of the two is closer to the lens and receives the image when the bird is in the air and the cornea is refracting. When the bird plunges into the water for a fish, the focal point shifts to the temporal or more distant fovea, bringing the prey into focus once more. A minnow of the genus *Anableps* utilizes both areas of keen vision at one time. As the fish drifts along the quiet surface of brackish waters, the iris is so shaped that the pupil is cut into a dorsal and ventral half, the former riding out of the water and the latter beneath the surface. Images in the air are received through the dorsal pupil on the lower half of the retina; objects below the surface are seen through the ventral pupil by the upper part of the retina.

The fourth mechanism is simply a highly mobile pupil that can be closed to a tiny pinpoint in the daylight. The "depth of focus" is thereby increased, giving a clear picture at all distances like the old-time box camera. Under water, the dilated pupils of seals, sea lions, and walruses focus normally on the retina. In the air, or in bright light, the tiny pupil compensates for the additional refraction of the cornea, and the animal sees fairly well.

Some vertebrates, like frogs and crocodiles, inhabit both media but see well only in one. These two groups are hopelessly farsighted under water but see well in the air. The penguins, on the other hand, see well under water but are nearsighted in the air.

Another problem faces animals that must see well into one medium from the other. A fish looking into the air at a fly—like the osprey looking down at a fish—must overcome the confusion of bending light rays. Their prey is not where it seems to be for the same reason that a stick that you poke in the water seems to bend. The discrepancy in the apparent location of the osprey's intended victim is obvious to anyone who has tried to shoot fish with a rifle. The fish, as it looks out of the water at an angle, faces a problem involving trigonometry and the refractive indices of both media. Moreover, the angle subtended by the surface of the water over a fish, through which it can see, is about 97 degrees. Outside that angle the fish sees nothing but the reflected bottom. Crammed through that 97-degree window is the whole bowl of the sky from shoreline to shoreline. The error involved in determining position has a sliding value according to the object's relative position above the fish. Only a fly directly over the fish is seen in its true position, and the same is true of the fish below an osprey. Probably both must strike from directly below or above to make a successful catch. The value of mobility is again emphasized in predation—and the eyes to go with it.

An extreme example of the problem is faced by the archerfish, a South American perch that shoots a jet of water from its mouth—knocking insects from shore vegetation or emergent weeds. The fish's eyes are under water, its mouth above. The shot is usually at an angle. If the surface of the water is choppy, the problem becomes even more complicated and would seem almost impossible if the wind sways the weed bearing the prey.

When we consider all these factors related to eyes and seeing, we come to have a deep regard for nature's ingenuity in providing its creatures with eyes for various special needs so that they may survive.

The Arboreal Eye

THE THERIODONT REPTILES and their ancestors carried it around for a very long time, having gotten it from some labyrinthodont amphibians toward the end of the Paleozoic. But it was still a sea eye. The lens and accommodation system had been modified and some external elements added to protect it in the air. Yet it was essentially the same one-chambered globe, good for both day and night sight, limited in color perception to the range of the spectrum that penetrates the ocean. The reptiles passed it on to the ancestral mammals, in whose keeping it retrogressed. In time a tree shrew of daylight habits rescued it from the dark and for the next fifty million years it became so intimately associated with a succession of monkeys in the splendid sea of the tropical forest that it may be regarded as an arboreal eye by virtue of the brain which it made possible and to which it was attached.

At that time some heavy-bodied monkeys, too large to keep their cradles in the treetops, strong and socially organized enough to fend off most predators, carried the eye and its brain back to the ground. It took nearly twenty million more years before they could write anything about what had happened. Then, having entered temperate climates, invented civilization, tamed their environment, and struggled through the ice ages (not necessarily in that order), it was easy for them to suppose that the old sea eye and its mentality belonged to someone else. The truth was that not much more had happened to it or the brain since those events began.

In the canopy of the hot forests the sea eye had been returned finally to vision in depths, becoming associated with an intellect more akin to that of the porpoise than to that of horse or dog. Wafted in wild frolic through the foliage, it had become a fast-moving or proto-tourist eye, preadapted to the pursuit of woolly mammoths at a later time and to broken-field running. In this flurry the imagery projected upon the curved inner surface is furiously skewed. Objects seem to grow and shrink in a graceful parallactic stream. It is surprising that this gradient of apparent motion filling the visual field does not produce a whirling, bleary smudge. But instead of shearing into fragments, the arborescent mesh of tree limbs that we pass seems to untangle, each twig taking its proper place in space. To a nearly stationary observer, such as a starfish, a scenery-hound with a camera, or a flatworm, all objects seem to extend at right angles from the line of vision, as they do in a painting. That they do not is constantly and delightfully disproved in motion, which permits the perception of distance as gradients of displacement of objects combined with gradients of texture, in gothic cathedral as well as jungle. It is not surprising that any monkey at the center of such a non-Euclidean geometry might feel himself to be at the center of the universe. The eagle is not more egocentric, despite his haughty expression, because more of his world is the clear sky which refuses to conform as a finite vestment around the flier.

The retina does not deal uniformly with this moving image. Its periphery is composed mostly of rod receptor cells, transmitting little clarity and no color. This part of the surface is oblique to the incoming light; the image flicking across it excites cells over a large area. Information from the "corner of the eye" is about motion: the silent approach of a hungry leopard, little restless movements of mates, or intention movements of one's competitors.

By contrast, the "central area" is part of the retina opposite the pupil. It is rich in cone receptor cells for bright light acuity and color vision. In it is a pit, the fovea, whose geometry and physiology make it the keenest point of a keen eye. To "look at" is to focus the image and the attention on it. Philosophers should give thanks for it. Without it they would be squirrels instead of monkeys. Their intellect would be no better than a good general alertness. The ultimate limits of philosophical speculation would be to isolate this or that in their thinking. In view of the primate reinvention of the fovea, which the older mammals had lost, the comic statue of the monkey meditating on Darwin's skull assumes a new poignant mutuality.

Foveate vision is not exclusive. It does not work like a telescope or a zoom lens, blacking out the periphery. It is an area of acuity and focus to which the whole visual field offers possibilities, and it remains part of the field. The mind is not seduced by what is at the center only, nor the fixated object entirely separated from the environment. A good historian, psychologist, or ecologist is true to his primate perspective when the scenery gives and takes on meaning relative to the object examined. Complete abstracting of the object is one of those "higher" intellectual activities that can produce new insights at the price of unimagined risk.

Color and central vision were part of the visual salvage operation at which most mammals failed. Most did not go into the trees. The ascent may have begun with following trails, nose and brain fixed on chemical litter and preoccupied with the world's delicious aromas. Perhaps the creature walked up the trunk of a tree like the animated cartoon figure who ambles distractedly up the side of a building. It extended the verticality of the quadruped's world, bringing the vertical and horizontal planes of life nearer par. These correspond to inner planes of balance and symmetry, linked to the gyroscopic activity of the inner ear and the sense of weight. The inner ear's canal system was devised in the sea for life in three dimensions. Modern artists, human and chimpanzee, know that the neo-sea eye follows vertical and horizontal lines, abstracting them from the profusion of visual experience. The oblique lines that primate art critics regard as having "tension" or "motion" are the leaning trees which are neither one nor the other in rectilinear space. From their armchairs the critics enjoy a heritage traceable to creatures who followed the lines literally but eventually discovered how to sit at leisure and follow the lines with their eyes. Even the most advanced terrestrial primates remain great kinesthetic tree-climbers.

During this transition the snout was traded for binocularity. It got in the way, obscured vision ahead, and carried unnecessary teeth and an increasingly obsolete sense of smell. Like the cats and owls, the ancestral monkeys refined distance-judging by bringing the eyes to the front of the head for stereoscopy. Then when they began jumping instead of merely scrambling, the horizontal paths, broken between trees, were completed by imagination, becoming gestalt figures.

The sloths, raccoons, opossums, and bears also went into the trees. But they did not jump and hence have to cope with such abstract thinking. The arboreal cats could jump and see well enough, but they committed the front feet to support. As meat eaters in the protein desert of the rain forest they remained spaced and solitary, which also closed the door to their becoming a terrestrial biped. Their very success as eaters of monkeys may have contributed greatly to the latter's socialization, however, as defense stimulated an elaboration of communication. The kangaroos and some other mammals did free the front limbs, but not to pick up odds and ends to monkey with or carry. They remained dull-witted, too, from a two-dimensional life.

But why did the tree squirrels not become monkeys—produce bipedal forms and terrestrial descendants? Certainly the rodents are a potent group, capable of enough intelligence, as in rats, to test psychologists. The answer may lie in their geography: They are temperate rather than tropical. Size is a prerequisite and the squirrels had to remain small enough to retire into the hollows of trees and climb to the ends of limbs. In a habitat of greater exposure they were often caught and eaten and had to produce about four new young per squirrel each year just to hold their own. This reduces the amount of education, tradition, and culture—a reduction that is not the route to humanity. Their emphasis on claws and jaws instead of fingers was related to a nut and bud diet. In spite of their genuine horticultural contributions of planting whole forests and their seasonal dependence on nuts, there has not been as much mutual evolution between nuts and squirrels as between fruits and monkeys. Squirrels lack a fovea and their color vision is not related to food finding. The paucity of brightly colored nuts may keep the squirrels from becoming human. Not that there is any evidence of disappointment. The first signs of culture among squirrels may in fact be an annual holiday for shouting "Hooray for the monochrome nuts!"

Conversely, color vision is beneficial to monkeys finding colored fruits in the tropical sea of green. It helps in differentiating all sorts of things in a world richer in kinds of things than the austere north. Foveate binocular sorting helps the monkey to know at a distance whether to eat things, run from them, mate with them, or examine them more closely. If curiosity killed the cat it was because he extended it beyond the permissible question of whether yonder moving object was edible. No doubt it kills the occasional monkey, too, as when a baboon cannot resist examination of a lion feigning sleep. But it put the monkey in touch with the extraordinary diversity of the wet tropics and excited him intellectually, as it did Alfred Russell Wallace and Charles Darwin. Color vision, sorting out, the grasping hand, intelligence, and leisure time combined in monkeys to produce a general plucking, picking, poking, and manipulating. They were learning to examine objects that seemed to be nonessential. Perhaps for the first time foveate vision was not merely a searching for signals to which there would be a stereotyped response, but an exploration at once aimless and more fertile.

The good eye for finding fruits was a good eye for finding fleas, too, and other sorts of mutual grooming such as the discovery of ethics, religion, and cocktail parties. All of these involve a new eye for the multiform expression of a peck order. Birds, by contrast, have great difficulty with ethics, social as many species are. This is not because they are simpler organisms but because their behavior is more rigidly inherent, and this poses the problem of how to have an ethical system without an alternative.

Flexible behavior implies learning, and learning is central to tradition. Tradition in primates involves a sense of social participation in a web of time. Flouncing through the canopy of the hot forests is a four-dimensional problem. It is not at all like loping across open plains where the future is visible in the distance and where direction is limited to movement on one plane. In the trees the surroundings stream by near at hand, all in the present. Seeing forward in space finds its means by seeing backward in time. The interchangeability of adjacent and successive order is a primate specialty. "I see" means "I understand"—that is, past events. A "seer" is one whose vision extends forward through the unity of time and space.

There is a self-consciousness about belonging to such a continuum It is embodied in the presence of the nose in the edge of the visual field. Though only a blur, it is a reassuring guarantee that the observer is not a disembodied being. Large noses remain an assertion of the self in a landscape. It does not mean the same to other animals, even big-snouted kinds. The opossum does not care much what it sees; the rabbit's vision is lateral. But among primates it objectifies the ego and justifies the nosiness of the prying eye.

When this sort of thinking was taken to the ground, the richness of its cerebral decoration was unpredictably pregnant. Much of it has become manifest in symbolic, metaphorical, and abstract perceptions and communications, the structure of which is modeled after the spatial interior of the arboreal world. The prehuman brush apes who gradually worked their way out of the jungle into the radiance of the open day never got so far that a tree did not seem welcome. There is yet an enduring nostalgia that is reinforced by the relentless glare and feeling of vulnerability in the open. An affinity for shade, for the nebulous glimmering of the pillared interior, for the tracery of branches against homogeneous backgrounds, for climbing, for the dizzy joy of looking down from a height, for hanging and swinging, for looking through windows and holes, for the mystery of the obscure, for the gothic fascination of interesting textures, for the bright reward of color—all these are part of the jungle past.

It was not possible to see the forest while in it, nor perhaps to create by planting. The temperate forests are not in form or impact like the hot forests. They are like dilute jungles with a cool openness and a remarkable geometry when leafless. The equatorial rain forest scarcely evokes comparable affection and is commonly said to have the appalling face of a formidable and implacable foe. But this judgment is only culture-deep. The tropical forest is profoundly more subjective and disturbing than a northern grove in the same way that the presence of an ape pierces our psyches more deeply than the presence of a dog. The ape evokes a relationship we feel in the bone, animating old, old fears and submerged joys.

REFERENCES

Cornish, Vaughn. *Scenery and the Sense of Sight.* Cambridge, Mass.: Harvard University Press, 1935.

Gibson, James J. *The Perception of the Visual World.* New York: Greenwood, 1975.

La Barre, Weston. *The Human Animal.* Chicago: University of Chicago Press, 1954.

Polyak, Stephen. *The Vertebrate Visual System.* Chicago: University of Chicago Press, 1957.

Shepard, Paul. "Eyes, Clues to Life Habits." *This is Nature.* ed. Richard W. Westwood. New York: Crowell, 1959.

Simons, Elwyn L. "The Early Relatives of Man," *Scientific American* 211(1) (July 1964): 50-62.

Walls, Gordon L. *The Vertebrate Eye and Its Adaptive Radiation.* New York: Hafner, 1963.

What the Eye Knows: Syllabus for a Course on the Evolution of Human Intelligence

If the self, or crux of identity-attention, is at the center of consciousness
and the capacity of that attention is based on intelligence, its outcome
knowing,
—then—

THE STUDY OF CONSCIOUSNESS EMBRACES

<u>What the "I" Knows</u>
but it begins, evolutionarily speaking, with

<u>What the "I" Nose</u>
that is, the ancestral, nocturnal prosimian, listening
and sniffing in a world temporarily structured from
sequential signals; there, in those Eocene jungles,
the "I" is the private respondent at the genetic-
environment interface . . .

and proceeds via arboreal leapers and specialized swingers
shifting groundwork to

<u>What the Eye Nose</u>
the diurnal simian, inventing spatially structured
order from sequential signals from the retina, but
thinking only "MonkeyMonkeyMonkey" and eternal
equilibration; here
the "I" is the implicit pronominal in the obsessive
social self . . .

but evolves, with bigger primates, to

<u>What the Eye Knows</u>
or the terrestrial strategist, for whom persistent
structure is signified by precise position and
eternal vigilance; to stalk or bolt, where
the "I" is the percipient subject-object in the
trophic structure . . .

and then to the tool carrier, burdened by

<u>What the "I" Knows</u>
this gaming, dimorphic symbolizer, coding his thoughts
and indulging in eidetic, vocal evoked, temporal displacement
(not the hear and now); and so, finally,
the "I" imposes and is predicated by the socio-ecological analogy.

 # Reverence for Life at Lambaréné

To THOSE INTERESTED in the human relationship to nature,
Albert Schweitzer holds a special place among philosophers. Human
relationships to God and to other humans have traditionally dominated the formal pursuit of wisdom. Schweitzer writes: "Ethics must
plunge into the adventure of making its adjustment with nature philosophy. . . . Let it dare, then, to accept the thought that self-devotion
must stretch out not simply to mankind but to all creation, and especially to all life in the world within the reach of man. Let it rise to
the conception that the relation of man to man is only an expression
of the relation in which he stands to all being and to the world in
general." He calls this ethic "Reverence for Life."

Unfortunately for those who hope that the ethic will heal the ancient separation and impel harmony in the human/nature community by providing for it a theoretical basis, it cannot succeed in practice. As a doctrine it may fulfill a strictly religious affirmation, and it may sensitize thinking persons to nature. But it is doubtful that many field naturalists would agree that, as applied by Schweitzer himself, it is ecologically valid. Reverence for Life does not provide a working pattern that conserves the natural environment of which particular kinds of living things are part.

The Origins of the Ethic

Ostensibly, Reverence for Life is universally applicable. But it has a history. It was born out of a religious Christian probing in the fertile field of Hindu theology and was nurtured on experience in Africa. The Indian ethic of no killing is a fanatic asceticism linked to impoverished landscapes that are already overhumanized and deteriorated. Applied in terms of a European ethos to the exotic African realm, its failure simply underscores the fact that one's relationship to nature operates through a particular culture.

Schweitzer's African writings mention many wild animals, and yet, in the light of Reverence for Life, almost every paragraph has some unexpected angularity. "A host of crickets began to chirp, accompanying the chorale that drifted over to us. I sat on a trunk and listened, deeply stirred. At that moment a hateful shadow crept down the wall. I looked up in alarm and saw a huge spider. It was much bigger than the most magnificent ones I had seen in Europe. After an exciting hunt it was killed." Despite his acknowledged influence of the ahimsa commandment from Jainism, the Indian doctrine, it is apparent that to Schweitzer all wild animals are not equally revered.

A swarm of tiny ants crowded around a drop of grapefruit juice on the floor. "Look at my ants," he is reported to have exclaimed, "just like cows around a pond." But on another occasion, columns of army ants passing through the settlement were hunted down with Lysol. "Thousands of corpses lie in the puddles," he wrote. "Serious enemies are the notorious traveler ants. . . . The militarism of the jungle almost bears comparison with that in Europe." It is evident that the same kinds of animals are not equally revered in all situations. The biologically indefensible comparison of jungle and European militarism suggests that to Schweitzer, who has seen so much human sickness and pain there, the jungle is essentially hostile.

The larger wild mammals invading the settlement are usually destroyed. When the door of the chicken house was opened, "twenty-two hens lay dead on the ground, their breasts torn open. Only the leopard kills in this way, for his first desire is to get blood to drink. His victims were taken away. One of them was filled with strychnine and left lying before the door. Two hours later the leopard came back and devoured it. While the leopard was writhing in convulsions it was shot by Mr. Morel." In overcoming the obstacles before his outpost hospital, Schweitzer's predator control measures were severe. This particular method is singularly cruel. Behind the Reverence for Life ideal, necessary killing may be accepted. But what of the means by which it is done and the preventive measures taken to avoid situations in which it becomes necessary?

Dr. Schweitzer had several experiences with hippopotamuses, including minutes of uncertainty as he passed near them in a small boat. He regretfully contemplated the destruction of a local hippo should it wander near the garden. Human victims of attacks by hippos, leopards, elephants, and gorillas were treated at the hospital. In nearly every instance that Schweitzer describes, the patient had wounded or otherwise antagonized the animal. He writes of gorillas that "attack immediately when they are seen and have the hunter by the throat before he can make use of his gun. . . . Sometimes they are real monsters, two and a half meters high." In repeating this misinformation on both size and ferocity of the gorilla, Schweitzer reveals his own lack of interest and curiosity toward wildlife. His books on Africa are practically devoid of original observations on nature. In one place he discusses the whale harvest of the southern seas in terms of the size of the harvest and its value. Snakes are mentioned only as enemies. "Wild animals are not dying out," he avers, citing as evidence an alleged increase in the elephant population due to the withering of native trapping techniques.

These instances and others like them are enlarged by their contrast to those moments of tender concern when Schweitzer carefully removes some small insect from danger in the middle of the footpath. It is always with sorrow that he kills. Only those animals must be shot, he says, which are destructive to crops or otherwise harmful. His judgment is as swift and staunch as though he had a millennium of precedent.

Nature in Protective Custody

The wild animals that especially engage Albert Schweitzer's attention are tamed and caged. Pet gazelles in particular were confined near his living quarters in a small enclosure. His many comments on this captivity reveal a paternal protective feeling. This takes the form of vigorous opposition to hunting for sport. Of falconry, for example, he says, "Is there a friend of nature who really finds pleasure in the tragic spectacle of the abuse and killing of a weak bird by a strong one, and who, by training birds of prey, takes pains to see that this spectacle shall be offered him as frequently as possible? . . . Falconry, moreover, is commended because it appears in the guise of sport and love of nature. It has no just claim to be associated with either." Here again is a seemingly logical extension of Reverence for Life grotesquely illuminated by Dr. Schweitzer's own behavior. Who is the friend of nature, the falconer or the captor of caged gazelles? Of all animals falcon and gazelle are among the fastest, widest-ranging, freest-moving; they are superlative examples of the fleetness and beauty we associate with wildness. The health of any wild creature is impaired by confinement—which is inevitably squalid among closely fenced hoofed mammals. Although falconers can make a good argument for their sport as a bond between man and nature, many people will agree that Dr. Schweitzer's attitude toward hunting is valid. Its oddness comes from his use of cages and his "exciting hunt" of the spider and other vermin.

Though it is rapidly disappearing through the activity of man, the rain forest as a whole seems to have no meaning for him. Nor is Albert Schweitzer much interested in preserving native Africa. Yet its helpless forms come under his mercy as individuals, and it has animal consumers that seem to him to prey mercilessly upon the helpless. This division of African animals leads us to those European ideas that have molded the shape that Reverence for Life has taken. When he says, "It is good to maintain life and to promote life; it is evil to destroy life and to restrict life," he apparently does not mean promotion or destruction of life only by humans. This axiom is not meant to be an admonition against killing, but rather is to be the basis upon which man decides which living things are good and which are evil, with the obligation of protecting some and destroying others. This is the peculiar Western twist to those Hindu precepts.

If Dr. Schweitzer were a naturalist, his judgment upon animals might be more troubled. He does not seek patterns in nature that might serve as criteria in any such judgment. "The struggle for existence that takes place all around us among living creatures can never be—unless we have succumbed to thoughtlessness—a spectacle to watch with interest and delight, but always a painful one." If evolution has produced things of beauty it is nonetheless ethically repulsive. So "our own vocation is not to acquiesce in the cruelty of nature and even join in it but rather to set a limit to it so far as our influence reaches." If the modus operandi of Reverence for Life cannot be identified with natural processes, from what does it extend?

The Ethic of the Barnyard

The answer in Dr. Schweitzer's case is that it is implemented from the traditions of small farming in Europe. Here animals are divided into two groups, wild or tame, and two other groups of even more overriding importance, predator or prey. The nature philosophy of the European peasant is primarily an extension of his knowledge of the barnyard community. In this view wild nature is a chaotic frenzy of predator/prey interactions. But the barnyard and the wilderness are not separate identities. The Western concept of the natural environment is derived from humanized environments containing some wild and some domestic species that interact to form a man-dominated community.

The behavior of all the animal members of this community is understood from what can be seen in the barnyard. Moreover, this community is seen as an allegory of human social groups. Besides being sources of food and materials, these animals have assumed roles and values as in human traditions and culture, connotations so finely embedded that the culture itself must change before any new appreciation of their natural life is possible. Thus the images of wolf and bear, pig and duck, have overtones in social attitudes of suprabiological meaning. The allegory provides the prototype by which children in this tradition learn to perceive nature, as well as a vivid lesson on the nature of the human social group.

The rural environment and peasant background from which Dr. Schweitzer came involve pastoral enterprises and a diversity of self-sufficiency to which we remain sentimentally and spiritually, if not economically, attached. This is the world of Mother Goose and, to some extent, the Aesopian fable. Mother Goose is full of scheming and just rewards, barbaric brutality and vengeance following the coarse law of an eye for an eye. Its system of punishment and reward keeps foxes and little red hens in rigid perspective. But besides portraying personality traits in the drama of human society, it predisposes a judgment on members of the humanized community surrounding the barnyard. This overlapping of Aesopian and literal roles makes it nearly impossible to keep allegorical and biological identities separate—or to evaluate anew the fox and his animal relatives without likewise sympathizing with the evils symbolized by him. Although naturalists today identify several types of mutual interaction or symbiosis and recognize that the equilibrium of the community as a whole depends upon their simultaneous occurrence, the traditional European–American attitude toward animals remains deeply conditioned by the predator/prey relationship and is usually a misunderstanding of that relationship based on barnyard experience.

In foreign surroundings, encountering exotic animals that have no niche in the fables, the European is apt to apply this dominant criterion as he assumes lordship of the habitat.

The Transfer to the Jungle

A biographer reports that Dr. Schweitzer had a boyhood ambition to be a swineherd. Appropriately enough, among the most cherished pets at Lambaréné was Josephine, a boar. With dogs, chickens, and sheep it wandered freely at the station, even into church. After Josephine persistently rubbed against members of the congregation, Dr. Schweitzer wrote: "I proposed to the missionary who was in charge of the station that because of all this I should kill her." Shortly thereafter, the boar killed five hens and "was enticed into the hospital, tied up, and expeditiously and artistically slaughtered." This satisfaction in artistic slaughter is consistent with Reverence for Life, for pig slaughtering is something that may be done skillfully or poorly, a part of the orderly universe of peasant existence well within the Christian view of animals in the service of mankind.

Some domestic animals harmonize better than others with the total wild-domestic community around human habitation. Schweitzer wrote: "I used to cherish the hope that the natives would little by little come to raise milk goats in their villages," but there were too many cultural and insect obstacles. The spectacle of impoverished peoples in stripped landscapes, bound to a downward economic spiral associated with a goat economy, is a common syndrome in Mediterranean and some tropical countries. Northern European soil is fecund enough to survive in spite of goats. The extreme ecological danger of pressing this animal upon African landscapes already troubled with fire and cattle was unrecognized by Dr. Schweitzer and still by others.

Many of Dr. Schweitzer's writings on animals are collected in a volume called *The Animal World of Albert Schweitzer: Jungle Insight into Reverence for Life,* selected by Charles R. Joy (Beacon, 1951). Here he outlines his amendments to conventional Western philosophy. He criticizes Kant for concerning himself only with man-to-man relations, attributing this myopic view to the influence of Descartes, who seemed to have "bewitched the whole of European philosophy with his declaration that animals were nothing but machines." He compares philosophers to the housewife who does not want animal tracks across her tidy floors. He observes that this dogmatic limitation on ethics persists despite the urgings of Schopenhauer and Stern, contrasting it to Chinese and Indian ethics. What Schweitzer does not seem to recognize is that animals have probably been left out of historical philosophy because there seemed to be no orderly patterns or fabric to their diverse ways. With no pattern to their interrelationships, how could they possibly be systematically embraced in epistemology? The naturalist's view today would be that, since Darwin and nineteenth-century naturalists, we do recognize at least the skeleton of community structure.

Let ethics rise, says Dr. Schweitzer, "to the conception that the relation of man to man is only an expression of the relation in which he stands to all being and to the world in general. . . . Ethics, therefore, consists in this: I feel a compulsion to extend to all the world around me the same reverence for life that I extend to my own." He admonishes men to break no leaf from the tree, pluck no flower, and to crush no insect with his feet. But how consistent is this with the statement given above that "our vocation" is not to "join in" nature but "rather to set a limit to it so far as our influence reaches"? The answer is that insects must be crushed by the virtuous in the same way that criminals are punished. The laws that it breaks are based upon human ethics that are extended "to all the world around."

In biological terminology this philosophy projects an intraspecific behavioral pattern (that is, human) as a standard for interspecies relationships of all kinds, with humans in judgment. This position places Schweitzer in a well-recognized class of people who believe that humans are here to impose order on the subhuman world. To the naturalist this application of Reverence for Life does not seem directed to the conditions and nature of animal existence, but stems from the exercise of mercy and compassion as purification of the human spirit. Put in the preceding terms, there is very little hope that such standards of ethics will successfully apply to natural communities anywhere except possibly in temperate Europe and Asia. There is evidence that the rapidity and degree by which invading Western man spoils his habitat is proportional to its difference from Europe.

By "spoils" is meant disruption of energy, material, and interspecific relationships, leading to soil deterioration, extinction of indigenous forms, and general lessening of the efficiency of the community as a whole.

The events at Lambaréné suggest that Reverence for Life in terms of the historical European orientation toward good and evil, and in the reflected image of temperate rural landscapes, collapses under its archaic burden of the hierarchic ladder of life. From the point of view of a naturalist who professes to be similarly sympathetic to the whole of life, it exhibits at least four distinct weaknesses.

The Shortcomings of Schweitzer's Ethic

First, Schweitzer's ethic slips into an oversimplified separation of sheep and goats, wolves and ducks, the good animals against the bad, the criteria being a confusion of traditional, symbolic, and economic attributes.

Second, it shows no cognizance of the life of populations or of the interdependence of species. Species, social groups within the species, and aggregations of species are real entities, though perhaps more abstract and less lovable than individuals. Nonetheless, the application of Reverence for Life is compartmentalized and preoccupied with individuals and single incidents.

Third, it is informed, at least partly, by hearsay, folklore, and the kind of spark-of-truth nonsense of the bestiary.

Fourth, it is motivated not by a genuine sympathy for animals or understanding of their own natural history and environment, but rather by a striving for self-purity and "self-devotion."

These criticisms, taken together, show that the criteria by which humans react to animal life have not been rationally identified. Animals do conflict with human needs, sometimes, and must be controlled. But the "preventive" principle upon which the army ants were killed is the same upon which other humans are killing millions of animals with chemical poisons, ignorant of eventual consequences beyond the temporary eradication of pests. The only difference is that Dr. Schweitzer expresses a profound feeling of sorrow. In the whole history of human war against pests, predators, and parasites there are numerous occasions in which the longer perspective has shown that complete victory is self-defeating and that obnoxious and destructive animals are essential in obscure ways to the health of the living community upon which humans are also dependent. The author of *Reverence for Life* has been astonishingly mute on the large-scale destruction of herds of antelopes and other animals in the African veldt in a futile and arbitrary slaughter to reduce the reservoir and incidence of sleeping sickness. In the end it proved pointless, accomplishing only the further decimation of Africa's incomparable heritage of animal life.

In fairness to Dr. Schweitzer it must be recognized that Reverence for Life has never dressed itself as an inductively derived scientific hypothesis. It is an article of faith, and as such it cannot be wholly judged on its consistency with natural history. As a facet of religious conviction it is invulnerable to the kind of criticism directed at it here. Its workability in this sense, its logic, its measure, cannot be decided by using field data as criteria. Even so, the Judeo-Christian tradition behind the basic outlook of Dr. Schweitzer does not absolutely separate religion and nature, nor faith from experience. Science is one of many human activities that may be called upon to inform faith. Because of their outlook, most naturalists should be very sympathetic to the general idea of Reverence for Life. But they will also feel that the expression it takes in practice, the way it is implemented and applied, must be consistent with their systematic knowledge derived from field data. The naturalists should propose to Dr. Schweitzer a modification of the practice of Reverence for Life emerging from the interaction of religious conviction and a broadened familiarity with the natural community. The value of this new affirmation would be enormous. For the first time, the naturalist and conservationist would see their knowledge of what needs to be done aligned with that body of faith which is the deepest source of our motivation as it enlarges the perspectives of philosophy.

As it is, Dr. Schweitzer's role at Lambaréné has been that of an autonomous father whose wisdom serves his compassion in adjudicating and healing. No one doubts the inspiring leadership of his mission. His compassion extends to all the natural world, and he rec-ognizes that our intellectual and philosophical substantiation of this feeling is weak. His remedy is to extend human justice into primeval patterns of animal life. Because this justice is ecologically naive and foreign, its effect is to destroy or alter those natural communities. This alteration, in the long view, may be detrimental to humanity.

Perhaps no European has more convincingly identified and crit-icized the homocentricity of Western philosophy. But the difficulty is not merely one of omission. The remedy does not lie in turning the machinery of ethics as it is now understood upon the animal world. A successful solution will have to include understanding of the web of animal life and respect for its own native community laws.

 # A Theory of
the Value
of Hunting

T HE PROBLEM OF THE ETHICS of killing is especially acute to
the biologist. Along with others, he must satisfy the perplexing questions raised by his own conscience. In addition, he bears a responsibility, especially if he is a teacher, to those who seek his counsel or
take the measure of his convictions. What follows is an effort to discover a value in hunting beyond the ecological effects on populations
and with implications for the ethics of killing.

Philosophical opposition to hunting is an established element of
our intellectual life. Even among those devoted to conservation there
are opposing views. Killing animals for the meat industry or for scientific research can be rationalized to the satisfaction of all but a few,
but hunting for sport is frequently regarded as morally indefensible.

In teaching and in other interpretive fields the issue is perennial. Values and attitudes are being formed. It is apparent to me that some of my acquaintances class hunting with war and murder. These critics are frequently students of humanistic disciplines, have potent weapons in a broad literary knowledge, are articulate and very keen, as it were, in the slaughter of the advocates of hunting. In a recent debate in a national magazine, for instance, Joseph Wood Krutch carved up his hunter opponent and served him to the readers—a murderer steaming in his own juices.

How has hunting been defended? One position is that sporting activity in the field somehow prepares a person for a higher plane of conduct in human affairs. This approach became obsolete with the formality and noblesse oblige of an aristocratic social structure, and may never have been valid anyway. It has been held that the stalk promotes character, self-reliance, and initiative. But this Teddy Roosevelt effect is unprovable. The development of leadership does not necessarily depend on the taking of lives. Assertions are sometimes made about "instinctive needs" and vague primitive satisfactions and psychological releases. The sharpest opponents of hunting sometimes give the impression that they have not yet forgiven Darwin and Freud. To suggest that hunting has psychic or evolutionary values only infuriates them. Then there is the claim that the hunter is really concerned with an "excuse" to escape the roar and friction of civilization, to squeeze out of society's trammels for a few hours of recuperation. The outraged response, of course, is that hunting with a camera is equally rewarding and more uplifting. And finally there is the Faulkner and Hemingway approach in which hunting is a manipulation of symbols for proving one's virility or otherwise coping with the erosion in the modern world of the human personality.

These rationalizations deserve to be junked. None of them is valid. If the real value of hunting is to get a hearing, its spokesmen must insist on greater perspective by all concerned. The essential point must clearly be understood to turn on a broader philosophy of human nature. Opposition to hunting for sport has its accusing finger on the morality of the act of killing. The answer is not a matter of forcing the admission that we are all human bipedal carnivorous mammals, damned to kill, but consists in showing through anthropology, history, and the arts that the superb human mind operates in subtle ways in the search for an equilibrium between the polarities of Nature and God. To share in life is to participate in a traffic of energy and materials the ultimate origin of which is a mystery, but which has its immediate source in the bodies of plants and other animals. As a society, we may be in danger of losing sight of this fact. It is kept most vividly before us in hunting.

The condemnation of killing wild animals for sport extends from some very provincial and anthropocentric premises. It is only a biased opinion that death is the worst of natural events. It is part of the naive assumption that order in nature is epitomized by living objects rather than the complex flow patterns of which objects are temporary formations. This view leads to the assumption that carnivorous predation as a whole is evil. A noted exponent of this idea is Albert Schweitzer, the author of the phrase "Reverence for Life." Dr. Schweitzer, who does not believe in hunting for sport, has sprinkled his jungle writings with accounts of righteous killing of predators. Europeans and Americans have always persecuted predators: the big cats, eagles, wolves, bears.

Joseph Krutch condemns the hunter for killing and claims that the distinction between life and death is one of the most absolute boundaries we know. Yet students of biology today realize that the chain of life extends into the atomic as well as the cosmic universe— that the most satisfactory definitions and descriptions of life processes are in physical and chemical terms. The organic and the inorganic are mingled at once in the living body.

The traditional insistence upon the overwhelmingly tragic and unequivocal nature of death ignores the adaptive role of early death in most animal populations. It presumes naively that the landscape is a roomlike collection of animated furniture. In this view the dissolution of body and personality cannot contribute to the orderliness that is necessary in an intelligible world. But when death is recognized in broader perspective, as transformation in a larger system, it can be seen to be an essential aspect of elegant patterns that are orderly as well as beautiful. Nowhere better than in ecology can we see that, without death, the elaborate and efficient natural community could not exist. The extremely complicated structure of this community has yet to be fully explored, but it should be noted that it is best describable in terms of events that constitute a field pattern. Plants and animals participate in it without questions (that is, sinlessly) in an attitude of acceptance which in human terms would be called faith.

The unfortunate misapplication of Darwin's theory to economic and class warfare in the late nineteenth century can still be seen in the reluctance to accept evolution as a significant factor in man's higher as well as his more primitive activities. Evolutionary theory also had the curious effect on some people of making nature seem more instead of less chaotic, especially with Spencer's unfortunate emphasis on conflict and the "survival of the fittest." As a process evolution is unrelated to the fate of individuals. One way to confuse our understanding of the means by which organisms have become what they are is to project our notions of ethics and our terror of death into our perception of them. There is nothing gruesome in the fate of animals who die before they have lived out their potential breeding or life span; this short span is characteristic of the natural world and essential to our understanding of populations.

The problem is not only acceptance of death as necessary to a larger order—and therefore greater value—than ourselves, but acceptance of killing. The moral criterion upon which humans kill plants and other animals is usually considered to limit killing to the necessity for food and defense. Under primitive conditions this meant something quite different from the modern slaughterhouse or the broad-scale application of chemical pesticides. The events of daily life in a hunting society are permeated with universal significance. Mundane behavior associated with fundamental requirements of life are not regarded as "merely" physical, but activated by unseen spirits. While we cannot expect our society to adopt animistic superstition and awe bordering on dread, we are tempted to admire, from an ecological point of view, the poignant sense of the interpenetration of persons and nature that they embody. We do not envy or strive to return to primitive ways, but we must acknowledge that their "reverence for life" is more selfless, more reverent, and more ecological than a fanatic overemphasis on death. To our highly sophisticated repugnance for soil ("dirt"), parasitism ("morbidity"), and decay ("slime") we add predation. We would prohibit it (as barbaric) from the natural landscape, extending remedial human "justice" into biotic realms where it is all but meaningless and incorporating democracy with its protection of the "weak" and containment of the "strong." Humans have become dominant in many habitats, but there is no process known by which this automatically extends democracy or any other political or moral system into the assorting of chromosomes, the adaptations of populations, or the interrelations of species.

Are we justified in influencing this tide of death? Do we assume a terrible responsibility in selecting the victims and killing them? Besides being murderous, are we not interfering in those natural patterns and upsetting nature's balance? Such questions reveal the initial false premise that we are demigods operating above and outside nature.

The taking of a life, so unimportant in a cosmic scheme, is nonetheless a profound event in our individual experience. Killing an animal probably does obliterate an individual awareness somewhat similar to our own consciousness. As sympathetic and vulnerable humans, we are sobered, moved, stunned by the death of any creature. This is why the tension over killing is so incisive and urgent. On one hand we have the cultural and personal necessity for tangible signs of our relationship to large-scale processes fulfilled in a moment of supreme excitement, and on the other our sympathy for a fellow creature felt most intensely at the crucial moment of death. An acceptable theory in defense of hunting must resolve these separate experiences and account for their unity.

So the hunters' apologist must ask for a more inquiring attitude. What does the human mind itself tell us of the hunter? Psychic well-being is associated with a mode of cultural behavior, and culture may be regarded as an interface between humans and their environment. Collective dreams, myths, and symbols, such as language, change slowly with the healthy functioning of society and the mental security of its members. There is evidence in literary and pictorial arts of an iconography of hunting. The complexity of the subject is a warning that hunting is not necessarily a vestige of barbarism or in form a wanton act. What is behind the word, "venery"? It is an archaic term meaning both sexual pursuit and hunting game. What is the historical relationship between the organization of sexuality and that of hunting? Is the value of hunting influenced by the human sex ratio or social patterns?

Hunting may be inherent behavior, but it is not just a behavioral predisposition. It is a framework of social organization that acknowledges an extrahuman context. This does not mean that killing is justified as indigenous or venerable. But it is a historical part of the activity of a people and should be regarded as having a place in the total fabric of what they have become and as a mode of their relationship to nature.

We are presently the dominant animal on much of the terrestrial earth. The stability of our own society is related to that of the natural communities of which we are part. If hunting promotes equilibrium within society it may benefit the stability of the natural community—which includes most of the plants and animals of the humanized landscape. These interrelationships remain to be explored in psychology, cultural anthropology, and ecology.

For perhaps ninety-five percent of our history we have been primarily hunters. Not much is actually known about this past. Therefore, living primitive groups supply information that may be taken as suggestive if not equivalent to past situations. The overwhelming evidence from hunting procedure among primitive peoples is that its execution is highly ritualized—a significant facet of religious worship. That we no longer worship animal gods is scarcely reason for condemning as pointless the subtle forms of the hunt. Interestingly, the exceptions to this tendency for the hunt to assume a formal style are found in areas of collision between hunting societies and civilized technology, where cultural deterioration has led to the breakdown of customs and to wanton killing.

Probably the richest collection of the ceremonies of propitiation of wild spirits by hunters before, during, and after the hunt is Sir James G. Frazer's *The Golden Bough*. Although perhaps anthropologically obsolete as method, Frazer's perspective and collecting genius remain monumental. To judge from the many examples he gives, hunting has been almost universally associated with ceremonial preparation and epilogue. When British Columbian Lillooet Indians dispose of the bones of their kill in a certain way—saying "See! I treat you respectfully. Nothing shall defile you! May I be successful in hunting and trapping!"—they are not just seeking to perpetuate their food supply. These rituals seem to solicit spiritual acquiescence and success. But such an interpretation may be the fault of our own suppositions. The ceremony makes less distinction between subject and object than we assume in the orthodox sense of magic. Frazer's view of ritual as coercive and petitionary magic was perhaps too restrictive. It also has a strong element of affirmation and communal participation. It was not only manipulative but also attuning, assimilative, and confrontative. Imitative magic is prototechnological and prescientific, but that part emphasizing "we-hood" and participation in a larger whole is religious. The magic and religion in primitive ritual reveal fundamental components in the hunter's attitude. The organized ceremony simultaneously serves not only a magic and a religious purpose but an ecological and social function also. It is aimed at maintaining equilibrium in the total situation. The whole of life, corporeal and spiritual, is to be affected.

The prey, or parts of it, are killed ritually and eaten sacramentally. By following the prescribed style, the hunters sacrifice the prey in evocation of events too profound for understanding. By its own self-imposed limitations the ritual hunt embodies renunciation in favor of a larger context of interrelationship. If the preliminary solicitation is effective and the traditional procedure is followed, the hunt is successful. Unlike farmers who must labor in the fields and who earn by their sweat a grudging security in nature, the primitive hunter gets "something for nothing." The kill is a gift. Its bestowal depends on the conduct of the hunters. Without this gift the hunter will die. As

Malinowski says, food is the main link between people and their surroundings and "by receiving it [they feel] the forces of destiny and providence." Of all foods meat is the gift par excellence because shortage of protein—not shortage of food per se—is the essence of starvation. The elusiveness of the quarry explicitly symbolizes the continuing dependence of human life on powers beyond human control. Hunting provides the logical nucleus for the evolution of communal life with its celebrations of a biosocial participation mystique.

What do the hunt and kill actually do for the hunter? They confirm his continuity with the dynamic life of animal populations, his role in the complicated cycles of elements, in the sweep of evolution, and in the patterns of the flow of energy.

It may at first seem irrelevant to seek present values for us in the strongly schematized hunting behavior of primitive peoples. On the other hand, Richard Chase says that we have the same basic needs as primitive man: "Our deepest experience, needs, and aspirations are the same, as surely as the crucial biological and psychic transitions occur in the life of every human being and force culture to take account of them in aesthetic forms." Many anthropologists report that there is widespread belief in the "immortality" of the spirits of all living things—a point of view we may not be advanced enough to share. Frazer wrote about forty years ago: "If I am right in thus interpreting the thought of primitive man, the savage view of the nature of life singularly resembles the modern scientific doctrine of the conservation of energy." The idea of organic interrelationship that ecologists explore as an inductive science may spring not from science at all but from a fundamental human attitude toward the landscape. In these terms, the hunt is a singular expression of our identity with natural processes and is carried out with veneration appropriate to the mystery of those events.

This concept transcends particular economic situations. In all sorts of societies—primitive, pastoral, agricultural, and technological—hunting continues fervently. The hunt has ceased to be the main source of food, but it remains the essence of a larger transaction. The prey symbolizes that which is received—whether from a host of animal gods, from an arbitrary god, or from the law of probability.

It is sometimes said that hunters are cruel, insensitive, and barbaric. In fact, however, the hunter may experience life more deeply. In a poem called "Castles and Distances" Richard Wilbur writes:

> Oh, it is hunters alone
> Regret the beastly pain, it is they who love the foe
> That quarries out their force, and every arrow
> Is feathered soft with wishes to atone;
> Even the surest sword in sorrow
> Bleeds for its spoiling blow.
> Sometimes, as one can see
> Carved at Amboise in a high relief, on the lintel stone
> Of the castle chapel, hunters have strangely come
> To a mild close of the chase, bending the knee
> To see from the brow bone
> Of the hounded stag a cross
> Grow, and the eyes clear with grace . . .

In urban and technological situations hunting continues to put leisure classes in close touch with nature and to provoke the study of natural history and to nourish the idea of conservation. Even royalty is subject to the uncertainty of the gift. From the Middle Ages we have numerous examples of the values of the hunt, where its forms have provided a significant social structure in complex royal households and its practice has stimulated firsthand observation at a time when hearsay and past authority were the main sources of information. The unique work of Frederick II in thirteenth-century ornithology is an example—an advance in our understanding of birds gathered in the course of hunting trips afield. A more recent example is the work of the late Aldo Leopold. A hunter and forester, Leopold documents in his books the slow sensitizing of a man to his environment through the medium of gun and dog. He postulates a "split rail value" for hunting—a reenactment of past conditions when our contact with the natural environment and the virtues of this contact were less obscured by the conditions of modern urban life.

Our civilization has extended the distribution channels of food and energy and has improved their storage to buffer lean years. The ultimate origin of food in the soil is no longer apparent to the average person. In this highly engineered and insulated atmosphere the natural world has become a peripheral relic, a strange and sometimes entertaining, sometimes frightening, curiosity. What has become of the gift? It has receded from view except to those who seek it. They may be found in the open country trying their luck. By various arbitrary limitations, both behavioral and mechanical, the hunter evens out his technological advantage at the start. This peculiar assemblage of restraints—legal, ethical, and physical—constitutes sportsmanship, a contemporary ritual. The hunt is arbitrarily limited. The hunter brings to focus his whole physical and spiritual attention on the moment of the kill. He expects to eat the quarry although it is dietetically irrelevant.

It follows that hunting is not, as even hunters sometimes claim, just an excuse to get out of doors, to which killing is incidental. Killing and eating the prey are the most important things that hunters do. The successful hunt is a solemn event, and yet it is done in a spirit of joy. It puts modern man for a moment in vital rapport with a universe from which civilization tends to separate him in an illusion of superiority and independence. The natural environment will always be mysterious, evoking an awe to be shared among all men who take the trouble to see it. If modern sportsmanship is a shallow substitute for the complex mythology or unifying ceremony of other cultures, some reasons are apparent: Only a part of the society hunts; and the ritual forms of this technological era are still young and poorly defined.

Regardless of technological advance, we remain part of nature and dependent on it. The necessity of signifying and recognizing this relationship remains. Hunters are our agents of awareness. They are not only observers but participants and receivers. They know that we are members of a natural community and that the processes of nature will never become so well understood that faith will cease to be important.

REFERENCES

Chase, Richard. *Quest for Myth.* Baton Rouge: Louisiana State University Press, 1949.

Frazer, Sir James B. *The Golden Bough: Spirits of Corn and of the Wild.* Vol. 2. London: Macmillan, 1920.

Krutch, Joseph Wood. "Sportsman or the Predator." *Saturday Review of Literature* 40 (1957):8–10.

———. *The Great Chain of Life.* Cambridge, Mass.: Houghton Mifflin, 1957.

Leopold, Aldo. *Sand County Almanac.* New York: Oxford University Press, 1949.

Malinowski, Bronislaw. *Magic, Science, and Religion.* New York: Doubleday, 1954.

Schweitzer, Albert. *The Animal World of Albert Schweitzer.* Translated and edited by Charles R. Joy. Boston: Beacon, 1951.

Shepard, Paul, Jr. "Reverence for Life at Lambaréné." *Landscape* 8(2) (Winter 1958–1959):26–29.

———. "Habitat and Painter: Ecological Process in Some Landscape Paintings." Paper presented to the Human Ecology Meeting of AIBS, Bloomington, Indiana, 1958.

Wilbur, Richard. *Ceremony and Other Poems.* New York: Harcourt Brace, 1948.

 # Aggression
and the Hunt:
The Tender
Carnivore

I N A CONFERENCE on "The Role of Aggression in Group Formation" the naturalist Konrad Lorenz commented that birds which
flock in a strange country fight in their home region. Physiologically,
the same hormone produces both fighting and courting—a common
observation in experiments in which hormones are injected into
birds during the season when they are not normally secreted.
Although the function of the nesting territory is reproductive, its
establishment requires an aggressive attitude. Particularly it has been
observed in certain kinds of geese that the touching display of mutual
feelings in the "marriage" ceremony is made against a background of
conflict.

A gander's territory is established by chasing out all trespassers.

The trespassers are at a psychological disadvantage when not in their own territory. If a gander looks up from one side of his property to see a strange bird, his endocrinal and nervous alarms go off and he attacks. A strange gander usually flees after making token resistance, but a female does not respond defensively at all. The gander swerves past her, rushing on to the far boundary where he briefly attacks his neighbor instead. It is essential in this drama to have a hostile and unyielding neighbor. The gander then turns back and engages the new female in a mutual posturing, a "triumphal ceremony." This is the first of a long sequence of behaviors that will cement the bond between them. The gander's aggression was first redirected and then rebounded into a stereotyped conjugal ritual.

Aggression is central to animal life. It occurs throughout the natural world, giving order and structure to the social web. As some animals have great strength, the problem of the control of power is widespread. Opposite is the tenderness that makes cooperation possible. When we contrive to exploit the efficiency of power but limit its misuse, we are engaging in a universal political struggle set deep in biological reality.

Pugnacity is patterned in animals. It is seldom the blind rage that we associate with "brute" violence. When it repels conflicting males it helps to space them into territories, each domain an insurance of adequate nourishment for the young. It helps spread populations into new habitats as the young move out to occupy territories of their own. The aggressive temperament impels the meat eaters, giving them the viciousness to catch and kill their food.

Fighting among the "gentle" herbivores, such as sheep, deer and rabbits, often involves disputes over territories or females. Like the fighting among birds, it is formal and ritualized. It is beautifully art-like, with distinct styles in different species, a dramatic dance like the fighting of Japanese samurai. Normally there is little injury. But in any situation in which the weaker combatant cannot escape, the fight between herbivores can lead to death as the stylized weapons become part of a singsong of grisly destruction. The relentless goring, kicking, or biting are not subject to inhibitory controls. The victor proceeds to batter his opponent with a horrible monomania, grinding him down like a driverless bulldozer.

Mutual destruction would seem even more likely among the carnivores. For them the problem of power and its misuse is always acute. In their normal activities, the large meat eaters must run down and kill another animal every day or so. They have become the world's most aggressive animals. These are members of the animal order Carnivora—the dogs, cats, bears, seals, weasels—and a variety of other species in different orders, including man in the order Primates. Among these predators aggression is not only a habit; it is an anatomy, a physiology, a behavioral pattern reflected in the whole metabolism and psyche. No such animal can be convinced that it should not be aggressive. Its viciousness is as much a part of it as its fur.

The prey of those meat eaters who hunt alone is usually smaller than themselves. To tackle large, powerful prey such as hoofed animals, predators live and hunt in groups. Despite their ferocity, their competition with one another, and their potentially mortal clashes, they cooperate, give mutual care, gently rear young and tolerate the young of others, and distribute food among themselves. Their pugnacity, indispensable to life, contrasts sharply with their tender complicity. The social hunters, such as wolves and lions, must resolve conflicts without slaughtering one another. When they are not actually cooperating to kill, what is to control their fiery tempers and disarm the coiled spring of savagery? A moment of discord may trigger their formidable muscles and weapons against one another. The males are always larger and stronger than the females. Thus the females and the young and the very old are in danger whenever a dominant male is present, especially if he returns home frustrated by failure.

In the daily nonreproductive activities the meat eaters, like other animals, have territorial and competitive disputes. The danger of these is partly offset by a system of social ranking initiated during the rough play of adolescence, forging an intimate awareness of personal status. Low-ranking individuals give way before threats by higher-ranking individuals. But the lower press their luck. Members of the group are always growing, changing, declining. Strangers arrive from the out-side. Fighting does occur. It also is stopped. Cowed wolves or dogs, taking the worst in a fight to decide rank in the social order or over a piece of meat, lie on their sides with one leg lifted, a position of exaggerated vulnerability. They could passively have avoided the clash, in the manner of the goose, by refusal to take a defensive atti-tude. Having committed themselves, they still have recourse to an "appeasement" posture. This is as nearly opposite in stance from intimidation as possible. Such submission postures are irresistible sig-nals to aggressors, who are immediately inhibited. Their performance and recognition are inherent, not learned.

We might wonder what the aggressor dog feels at this moment. The emotional experience probably changes abruptly from rage to something else. We can only guess how this feels. Possibly one would say: "Doggone, that happened to me only last week!" But more likely it would be: "Why, I don't really believe I want to tear you apart. I despise you. Then again, I feel rather sorry for you." From such hum-ble beginnings in conflict spring the ties of affection and tolerance.

It is no accident that the dog is our most consistent nonhuman companion. The circumstances of life among wild dogs are surpris-ingly similar to those faced during thousands of generations as hunters, when we lived by killing horses, bears, bison, cattle, deer, mammoths, giant sloths, and other large mammals. Certain adapta-tions, such as the use of the hands for tools, were impossible for quadrupeds; others, such as the development of large canine teeth, abandoned by humans, were developed by the dogs. But the problem of coping socially with savagery was met by humans and dogs through inhibitory and appeasement expressions and postures, a spe-cial repertoire of domestic anodynes. Male dogs never bite females. Their intense ferocity is counterbalanced by sweet loyalty. As dog-lovers know, dogs are capable of great depths and empathic displays of affection. Because they are social hunters—tender killers—dogs could be grafted onto human society or humans onto dog society. This could not happen with bears, even though they look more like men than dogs do, because they hunt alone. They are partially vegetarian and capable of monomaniacal murderousness. Humans adopted bears into human society largely in mythological ways rather than by domestication. Among the cats, by contrast, there is Elsa, the Adamson's tame lioness, who proved so charmingly that the social great cats can join the human group in an affectionate symbiosis.

Homo sapiens descended from certain Miocene apes who presumably had arboreal ancestors. They were omnivores, not carnivores; gatherers, not hunters. Even now the human digestive system, tooth structure, and certain other parts remain as typical of plant eaters as animal eaters.

So long as they remained omnivorous, prehumans were more like the present-day apes than like ourselves. The opportunity for change probably came with climatic revolution. Prehumans remained stuck with the picayune food sources of the tropical forests so long as the environment was wet and hot. They may have become incipient hunters from seed-gathering or opportunistic hunting in tropical savannas. Possibly as a consequence of momentous mountain building by earth crustal movements, which reached a climax during the late Pliocene, the climates of much of the world changed and the great grasslands emerged with their herds of grazing animals. Loren Eiseley has said that if it were not for the grasslands we would still be sitting in the dark munching roaches. So came the discovery and challenge of hunting, of cold and aridity, and of social being as a carnivore. Of all the great revolutions of human history—the invention of agriculture, the discovery of metals, the creation of cities—none has been more fundamental in every dimension of human shaping, producing the supremely human capacity for aggression and love.

Arboreal ancestors bequeathed us hands, an emphasis on braininess and sociality, closer mother/child contact, prolonged development of the individual, and a modification of the sexual cycle away from an annual season. Possibly only a creature with these potent endowments could have so refined the biological beginnings of love. *Australopithecus* in the early Pleistocene and later *Pithecanthropus,* both prehumans, mark states in the transition from gathering and incidental hunting to the perfection of organized hunting of large animals.

With the help of fire tools the australopithecines or their descendants created a new kind of society based on food sharing and symbolic communication. Food sharing on a year-round basis is unusual in nature. It is especially efficient where food comes in occasional large concentrated protein packages, where distribution is important before it spoils or is stolen, and where fairly sophisticated cooperation and communication are necessary to obtain it.

It was a self-stimulating complex of events. The substitution of tools for canine teeth and fire for jaw muscles were part of the changes making a speaking mouth possible as part of the evolution of the face. In the primates and carnivores, apart from sound production, faces are flexible expressive signaling devices. Perhaps even among those cavemen, who probably founded the permanent domicile and human monogamy, an aesthetic selection for faces had begun, based on a choice of mates and the necessity of expressiveness upon which social peace depended. This train of emergence carried increased differences between male and female and separation of their roles. Greater strength and size of males were probably selected by competition and hunting; the morphology of females was probably selected by their progenitive role and the survival of the young.

Disarming the aggressive temperament was undertaken in humans, as it had been in other meat-eating animals, by ritualized inhibitory postures, by expressions of submission, and, eventually, by positive symbolic signals as life became more complicated and elaborate. Assault was deflected, modified into games and art, or ritualized and controlled. Coming down to us from the prehumans who invented these protective devices to turn the hunter's ferocity away from home and clan are many gestures in vestigial form, postures culturally elaborated and reworked. Thus the servile bow of the warrior-slave, the humble genuflex of the devoted, the vassal with hat in hand before his baron. Although derived from animal submission, these attitudes no longer deflect wrath so much as they confirm and betoken a bond, acknowledge a relationship; yet they probably originated as pleas for mercy, supplication by vincible individuals in which the vulnerable back of the head and neck were exposed. The knight kneeling before the king is neither timid nor weak; he kneels to dedicate his zeal and weapons. The "suffering servant" who voluntarily puts himself at the service of another symbolizes his humility on his knees by washing the feet of someone to whom his compassion is directed, a forgiveness for aggression and a posture drawing forth love in its stead.

The position of the woman in this carnivorous world was especially poignant. The hunting society both increased the man's aggressiveness and created the human family. The male was not only liable to flare up at the female, as toward his hunting companions, but his aggressiveness was related, as in geese, to his sexuality. He pursued the woman with an intensity akin to that with which he chased woolly mammoths. Venery means man's hunt of game and of woman; art and civilization retain the elaborate metaphors of pursuit. The hunt ends in death and so does love, as the death of the prey at the end of the chase echoes the inevitable death of all who complete the life cycle in love. Both kinds of prey are shot, one with Diana's arrow, one with Cupid's. Since the horse was domesticated and used for hunting it has been associated with the erotic content of the hunting imagery. From men on horseback the prey's body is pierced. The spear, arrow, gun are phallic and are incorporated into primitive fertility rituals.

It was necessary for the woman, whether the target of redirected fury or the sexual prey of the wild huntsman, not only to deflect the violence but to entertain it as well if pair bonding as well as community harmony were to develop. The first step of this delicate problem was solved long ago when the period of female sexual receptivity was extended and reproduction detached from an annual cycle. Then early humans invented special signals. The general hairlessness of the body may be related to the discovery of Eros-inciting signals that were emphasized by modified submission postures, refinements of a position called "presentation" by primatologists. Other bodily changes occurred. Breasts were maintained in form between lactations. Subtleties of face and eye expression were increased. The pubic triangle is probably hirsute for this reason. Humans are unusual among animals in the use of genitalia themselves as sexual signals.

In addition to these anatomical signals for channeling male wrath, woman adopted certain artifacts that inspire a libidinous impulse. These vary from culture to culture: jewelry, bound feet, bustles, hair arrangements, perfumes, tattooing, and much of the clothing. In extreme circumstances the most singularly arresting attitude the woman can take is to face the angry man naked except for her jewelry. Perhaps it is not surprising that hoodlums' molls are either over-dressed or have nothing on. They are the extreme response to some of the most aggressive men in society.

When the aggressor's violence is transformed we may, observing from the outside, say that the person behaves. We know from our own experience that this behavior has an affective or inner component that is more than erotic excitement. Tenderness is the emotional aspect of cooperation. What we *feel* is the heritage of a million years during which hunting and tenderness emerged together: a reworking of an even older primate heritage in the creation of a new and deeper sensitivity. It owes something to those mild vegetarian ground apes who brought the fine primate nervous system to terrestrial habitats and who explored and carried things and ran and threw. But we owe the prehistoric hunters, possibly the most aggressive animals ever to live, credit for the capacity of human love. In every species of animal the emotional range from hostility to tenderness may be like a tree—a tree whose high crown indicates a very deep taproot: a great depth of aggressiveness balanced by a lofty affection that is unmatched elsewhere in life.

This affection has an emergent quality in humans. The processes of turning our native ferocity and cultivating its opposite have become extremely complex and subtle. It is an axiom of Christianity that this should be a deliberate activity, and in our logical way of thinking we may suppose that conscious effort is the only way. But we extend an instinctive foundation which we share with the killer animals and to a lesser extent with all animals.

The modern hunter goes into the field with a new perspective, turning to the hunt proper with a new tenderness. The aggression/cooperation tension is most beautiful in the encounter of man and woman, however. Unlike ordinary daily interpersonal meetings, one of the two is a kind of prey. The hunter is pursuing affection. The prey, at whom the violence of a hunt is directed, is the same being in whose embrace are found the counteracting events indispensable to human survival. So do hatred and compassion, death and regeneration, hover inextricably over love.

 # Meditations
on Hunting

EMERGING FROM the Racquet and Tennis Club of New York, where as a guest I had been scanning books on hunting in the club's library, I looked down Park Avenue with its towering, monolithic buildings, its patina of wealth and power highlighted by elegant corporate foyers and suave inhabitants, and I was wrenched from my subject. How trivial seemed hunting against this overwhelming impression of infinitely complex machinery! I felt a sense of absurdity. Could anything in this glut of human success be more utterly meaningless, except perhaps in some recondite, academic way, than reminiscences by old duck shooters and manuals for young falconers? What could contrast more with the country pursuits of a few wealthy sportsmen or even the prehistoric arts of the chase than this awesome street that epitomized the will of man? All of nature and its study seemed remote. Only a faint yellow stain in the air hinted at some underlying disorder and the cost of such an imperious civilization.

The earth and air are strained by human dominance, and the bonds with nature are broken by the definition of history. Nor are those bonds ever self-evident. Scientific understanding of the interdependence of species has accumulated slowly with ecology's rather ponderous investigations of symbioses, successions, and communities. A half-century's work has uncovered a principle: elaborate and stable flow patterns linking forms through food habits. Such food relations—the food chains and food web—define evolutionary categories, curb and channel and separate ways of life. The lives of all creatures are shaped by the details of feeding, not only in their physical anatomy and chemistry but in the most precise aspects of form and behavior.

Among the biological characteristics of the human species is the specialization of some parts of the central nervous system for storing and transmitting information symbolically. The older views that mental and cultural life set man apart from nature or that it constituted a remission from evolutionary selection are challenged by recent anthropological work. Comparative studies of many different primates suggest parallels to the prehuman situation of our evolution and the steps by which social, intellectual, and even ethical traits came into existence in an ecological context associated with the human animal's niche and, more particularly, his place in Paleolithic food webs.

These studies are slowly delivering a picture of the human mind as an adaptation: to the physical environment, to band, clan, and tribal organization, to the division of labor in hunting and gathering, to long life and delayed maturity, to the ceremonial hunting of large, dangerous mammals living in herds in open country—in short, to an environment and a way of life, the life of the Pleistocene hunter.

Together with physical traits, these are perfections generated in a hunting/gathering milieu shaped during ninety-nine percent of human time. Natural selection has directly created the most subtle and delicate aspects of thought, passion, and art. We have gradually accepted human hands and legs as those of a hunter. Now we are ready to find in our heads the mind of the hunter: the development of human memory as a connecting transformer between time and space, derived from the movement of hunter and gatherer through a landscape; the Dionysian moment of unity and freedom in ecstasy of intense release unknown to herbivores, based on the recognition of the perpetuation of life at the moment of the kill; the mental use of other species as metaphors for social perception, which is a mode of totemic transformation reaching to the heart of thought, yoked by our evolution to the richness of world life, ushered into consciousness by the magic of the taxonomies of the species system and the terminology of anatomy as revealed by the hunter as butcher.

These recent developments in bringing social and natural sciences together are astonishingly anticipated by José Ortega y Gasset in *Meditations on Hunting*. When Ortega calls the present-day poacher "municipal Paleolithic man," he does not mean that he is low and brutal. He has perceived by an almost incredible act of intuition into the thoughts of hunters he has known personally and from literature that the glory of man is a hunting heritage.

Ortega recognizes that the antiquity and universality of the hunt signify deep, positive traits. He knows that routine work and drudgery are an invention of civilization, degrading to the human spirit. He acknowledges the purposeful uselessness of hunting, the handicaps by which hunters always neutralize their technological advantage, so that the hunt is changed from what we commonly regard as strictly economic. He perceives the ecological importance of size relationships among the hunters and hunted, which is the fundamental factor determining the number of links in food chain systems. He spurns the deceptive substitutes for hunting truly. He ponders the behavior of prey and predator as necessary to each other and interlocking, and he knows that game cannot be saved. He speaks of the numbers of animals as related to their role in the prey/predator series. He sees that domesticated animals are degenerate. He affirms the hunting way of life by man as a combination of high technical ability and religious sophistication.

Throughout Ortega avoids the implicit full knowledge of the cliché that "Man is an animal, but something more than an animal, too." Instead, Ortega says, "Man has never really known what an animal is." In this he turns away from the orthodox Western attitude that the nature of animals is self-evident; that tradition condemns the carnivore and the act of killing as evil, and now and again it reaches to the Orient for its reverence for life. But Ortega knows that only the hunter confronts this question with full human dignity, beginning with an admiration of his ecology rather than its denial.

His description of the hunter is seldom abstract however. One of the most attractive aspects of his essay is the description of the individual hunter in the field: his minding of the environment, the fluid quality of his attention, and the habits of alertness and acuity that link him in participation with all of creation.

The hunter's vision is itself a part of nature. His perception of the signs of passage and signals of events is continuous with those events. His eye roves across a landscape which is itself living. The hunter lives an eventful life, a present, sound-filled pulse that collectively is the dynamic, oral, traditional society, where the poet is historian and humans are bound in myth and music to a generous and religious existence.

The static cornfield, the static steel towers on Park Avenue, the static nature of print itself are, in a way, paltry gains. Recent peoples are terrifyingly separated from the organic root, frozen in fear of the necessity of being consumed, sick in bodies and minds that have tragically outlasted the great Paleolithic religious institutions.

When hunting was a way of life, work as we know it did not exist. Since then only those fortunate enough to be free from work have kept the vestiges of hunting traditions. The great privilege of leisure has been hunting (and so it is not surprising after all to find a library of hunting books on Park Avenue).

It is a mystery to me how Ortega arrived at so sure a footing. He did not live to see the golden days of African anthropology, with fossil humans virtually springing from the earth like cicadas, or the scores of educated ape watchers who have been finding kinship beyond their dreams in the jungles, or the bitter maturity of ecology, which was little more than an intuition before a decade of fallout from nuclear bomb testing triggered worldwide pollution consciousness. Yet he anticipated with profound accuracy the direction and basic formulations of a discipline that does not yet exist: a true human ecology.

Ortega has grasped that essential human nature is inseparable from the hunting and killing of animals and that from this comes the most advanced aspects of human behavior. Intelligence in our species is a highly specialized function by which natural selection built the hunter's capacity to plan on the intensely social foundation of primate ancestry. The emergence of such complex functions as the unconscious, of language, and of dreams are inextricably part of the Paleolithic. Ortega does not scorn but affirms that past. A view of man so humble in the scale of nature and so audacious in its challenge to the homocentrism of our traditional philosophy will not be easily accepted. Yet like the idea of ecology—indeed, as part of ecology—it will in time reach out to all areas of concern and thought.

First, however, it must penetrate the husk of five thousand years of civilized fear and hostility toward nature in general and hunter/gatherers in particular. Only when our culture accepts the needs of living men as shaped by a prehistory that is still urgent in them, communicated to each by his chromosomes, will we be ready to follow the lead of Ortega's *Meditations on Hunting*.

 # The
Significance
of Bears

I N MY ONGOING ENCOUNTER with the ideas of bears, I regard myself as belonging to a noble stream of consciousness that links me with my ancestors—not those you dig up in the genealogical library but those whose bones are dug up by archaeologists. It is a connection also with the tattered, present-day survivors of the Paleolithic, those hunter/gatherer remnants of the Pleistocene, and the three million years during which our species differentiated from other hominids and achieved self-awareness—perhaps the ultimate human possession. The Pleistocene ended about ten thousand years ago, as the last ice age lifted its burden from the northern world. It preceded all agriculture and all towns.

What we lost with that wild, primal existence was a way of being for which the era of agriculture and civilization lacks counterpoise. Human life is the poorer for it. This loss is one whose details I have explored elsewhere. Here I wish to define it in terms of the bear and to identify it as a shift from the concept of life as generous and giving, as signified by the bear, to one in which gains are negotiated through the patronage of a willful god.

There are many ceremonies and myths in the human/bear relationship, but the most important are those associated with its killing. These represent ways of perceiving and thinking of the bear that mark the cusp of a whole cosmology, a kind of covenant between the animal world and the human world. This three-day to ten-day celebration among the circumpolar tribal peoples arises from concepts about the place of humans in the world. The bear, because of its unique charisma and physical value, happens to be the paramount agent provocateur or occasion in that relationship: the major spiritual affirmation of human life in accord with the way of nature.

The Paleolithic cosmology holds our existence to be framed by "a gifting world." It is particularly evident in the prayers of the American Indians of the Southwest—for example, in their numerous expressions of thanks and gratitude—and perhaps, in a faded way, in thankfulness in all religions. But it is not, in its pristine form, the end of a transaction in the sense that a merchant says "thank you" at the completion of a sale. And in this lies the fundamental difference between the old hunters' conception of a *bestowal* as opposed to a *claim*.

The world of the holy bear is a beneficent universe in which, typically, there is plenty. It is a beautiful world, to which we are given access by birth, just as each day is a new bestowal. Our only indemnity is to accord it and the divine messengers the reverence and gratitude they deserve. All that we have and are, in this ancient view, is freely given. It entails courtesy and obligations on our part, of which the ceremony of the slain bear and other hunting protocols are the most notable. In this Paleolithic philosophy, human society, judging from the evidence among its living survivors, incurs the duty to share food, especially meat, along family lines or according to the circumstances of the hunt. But in finding the bear or receiving a portion, one does not incur a debt. One does not haggle with the unseen "master of the forest" or the "master of animals" who is said by some tribal peoples to renew the animals. In its killing, the bear is never "taken" in the sense of human striving against it or conquered in the pride of victory. The hunters do not perform personal acts so much as the work of nature.

When killed, having been given to people by the forest divinity, the bear does not entirely die. Its spirit hovers nearby and its presence is felt throughout the days of the celebration. A going-away ceremony closes the festival: the bear's skull is placed in a tree, symbolizing the rejoining of all animals with the plants on which they ultimately depend and connecting the consummate animal with the great tree of life that links the heavens, the earth, and the underworld. Finally, the bear's spirit returns to its mountain source. Its eventual renewal and return in another season as another physical bear follows the fulfillment of the rites expiating its killing and the appropriate acts of veneration. But these are not a bargain with the spiritual world, however. They are a confirmation of the natural order.

Rites of sacrifice, by contrast, are a speculation in the power market. "Sacrifice" is sometimes misapplied to the bear rites. But sacrifice is a domesticated idea, invented by farmers and pastoralists and elaborated by tradesmen and their human overlords. A sacrificial offering, usually a sheep or a goat, is killed and offered to a deity at an altar as the propitiatory token of a radically different consciousness than that of hunters—a fundamentally altered view of humans in the universe and the powers with which they must deal. Most sacrifices are sacramental: centered around a sacred feast, a portion of which is dedicated to a god or goddess in expectation of a return. The offering is itself an immolation. A self-immolation lies at the root of Christianity, linking it to the post-Neolithic religions which, according to the records, all made religious sacrifices.

The motive for sacrifice comes from a sense of food as a commodity to be garnered, stored, and traded: a means of power. The sacrificial approach to religion recognizes food as a medium of exchange: a product of work. It conceives the character of the spiritual powers as fundamentally like that of humans in the sense that people are all entrepreneurs in a world of scarcity—where consummation follows exchange. Even the gods are greedy. They reign in the spiritual world much as nobility reigns in a world of taxes and centralized authority. In this sense the victim of a sacrifice is given as an investment.

Among the hunter's rites, a portion of the bear's own meat is shared with its ethereal immanence—with the bear itself. In this sharing there is no magical compulsion, no dealing. The idea of sacrificial offering by later generations is a corruption of the original act of gratitude transformed into an attempt to bribe transcendent powers—themselves regarded as greedy, touchy, querulous, jealous, and obstinate in granting favors. Indeed, sacrifice at an altar has unmistakable similarities to the appeasement of kings, nobles, priests, and others who are brought gifts in order to seek favors or atonement.

The distinction between gift and sacrifice may seem at first mainly of interest to anthropologists and other scholars. But the differences between, on the one hand, the northern hemispheric cult of the bear that is still practiced in remote parts of Eurasia and North America and, on the other hand, the worldwide practice of sacrifice, with votive offerings of plants and animals, along with the myth of a self-immolated messiah, reveal contrasts that are at the root of the psychology of being.

Sacrifice belongs to a world in which provision must be made: A withholding world is struggled against, stores are put by, and powers conciliated. Most of us belong to such a world where we bargain, if not with the spiritual rulers, then with the trades and costs; everything has a price. We "sacrifice" the hope for a new house in order to send children to college, see ourselves as "sacrificed" in war, and speak of animals in scientific laboratories as "sacrificed" for research.

It is not only a matter of chosen ideology. There is no longer room for the gifting philosophy because the world is too crowded and human wants too great. It can only be afforded by an "original affluent society" that has not yet exceeded the land's carrying capacity. It is beyond the hope of those whose priorities are the limitless accumulation of money, things, and human copiousness. In such a world there is inadequate space for the brown bear, whose biology demands unhindered room in wild country. The end of the bear as a physical presence coincides with the end of a human way of being at peace on the earth. The lack of bears and human disillusion are not merely coincidental: The bear is the gift par excellence of a generous cosmos, the occasion around which humans have recognized this benefaction and shaped their gratitude for the whole of life. It is as though the world had provided a model recipient and at the same time a supreme gift: the only other large omnivore in the same habitat as humankind.

The bear is a playful creature with a face and a mind, a creature who deliberately rears its young toward an end, attends to its surroundings in a most suggestive way, seems to put its faith in the renewal of springtime, and yet chooses the pathway to the human settlement. It goes calmly to sleep in the fall, as though confident of spring's bounty, and equally gives itself, according to the hunters' myth, to death at human hands. The "luck" of the hunter determines the discovery of the den—but hunting/gathering people know that luck is never chance. The bear is the supreme Gift, rewarding the people who play their part in the great round and acknowledging the bear's "voluntary" presence. Yet the bear does not sacrifice itself. Nor is it self-sacrificed to redeem humankind. Nor is obtaining bear meat and fur a confrontation with a ferocious beast. It is attended by quiet resolution and a blow to the head as it is awakened in hibernation.

In the human overabundance that accompanied civilization, we all became victims of the unequal distribution of wealth, necessitating a manipulative ideal and the role of sacrifice. Before that we had long believed that the upper world was a kind of equivalent of this world, populated with spirit plants, animals, and people, not a heavenly oligarchy of beings with human features and human flaws who could be plea-bargained. In recent centuries we have, like gods, taken the power over nature into our own hands and think we no longer need to roast a lamb at an altar to get a crop of corn. Even so, we are engaged instead in forcing the soil to yield, making the soil itself the victim, sacrificed to our inner gods in the same bartering spirit.

Looking to the bear will not restore me to those distant ancestors who preceded by hundreds of millennia all that negotiation and debasement of the spirit. But it may open my heart and mind to the double gift of the bear as feast and physician in its role as the killed and renewing deity whose grease, once tasted, is supremely relished over any other "fat of the land," and whose wildness reminds me of my wildness. The bear sustains me yet.

The bear gives physical sustenance and spiritual healing. Years ago I had delicious meals of bear meat and I cooked with its fat for many weeks. More recently, in search of health in New Mexico, I entered a native healer's house. In a firelit room he was ready in traditional regalia, surrounded by a rich array of paraphernalia. The ensuing smoke, teas, chants, dances, and songs washed over my senses. As the hours passed I drifted in the nexus between the physical body and the spiritual realm of medicine. I was aware of being embraced by black, hairy arms and hugged with paws with claws. In my ear was an unmistakable snuffling. The twofold gift of the bear was fulfilled.

PART II Searching for Place

PAUL SHEPARD'S INQUIRY into what might be called the "history of ecological ideas" began formally when he entered graduate studies at Yale University in 1950. His interest in nature aesthetics and nature perception in American settlers grew during the next twenty years to include the orientation to the world of the first humans, our hunter/gatherer ancestors. By the early 1970s his research had led him to speculations on "the impact of place" on human development and ecological understanding.

During early adulthood, Paul was very fortunate to be associated with an outstanding group of mentors and role models. His doctoral committee at Yale brought together Paul B. Sears (ecological conservation), Ralph Henry Gabriel (American history), G. Evelyn

Hutchinson (zoology and ecology), and William Jordy (art history)—all outstanding scholars who guided him through his interdisciplinary doctoral degree at Yale. While serving as conservation chairman for the National Council of State Garden Clubs he became a member of the newly constituted Natural Resources Council of America. In this position he found himself at the center of a remarkable band of environmental activists with whom he lobbied in Washington, D.C.: Dave Brower (Sierra Club), Charles Callison (National Wildlife Federation), Ira Gabrielson (Wildlife Management Institute), Joe Penfold (Izaak Walton League), Fred Packard (National Parks Association), and Olaus Murie, Sigurd Olson, and Howard Anhiser (Wilderness Society). With Rachel Carson he testified on the harmful use of pesticides. Undoubtedly the influence of these impressive environmentalists and naturalists was instrumental in shaping his own ecological consciousness.

Paul didn't leave his intellectual development to associations alone. He conducted ongoing research in libraries and journeyed to sites to see for himself what that research had uncovered. He followed the paths of the Hudson River painters and took photographs where they had set up their easels and romanticized the early American landscape. Traveling and studying abroad, he studied the formal gardens and Renaissance art that formed the aesthetic framework for the perceptions of settlers to this country. He studied the journals of travelers and paintings of artists who traveled the Oregon Trail and then followed their paths to the points repeatedly mentioned or depicted, again taking pictures and comparing them with old paintings and photographs.

As a young professor he put his knowledge to practice when he worked summers as a park ranger in Glacier, Crater Lake, and Olympic national parks. When his ecological vision came up against the stark reality of the use and misuse of public resources, he got involved in the fledgling attempts of environmentalists to stop uncontrolled exploitation of public resources. His summer employment in national parks came to a sudden halt, however, when he became a whistle-blower and key figure in uncovering illegal logging operations carried on in one of the parks. He was subsequently denied further employment with the National Park Service. Back at Knox College, with the help of students and faculty, he restored the farmland of Green Oaks to tallgrass prairie—the east prairie was recently named after him in honor of his accomplishments and dedication.

An extremely prolific writer, Paul produced a steady stream of essays, including groundbreaking work on the natural philosophy of gardens and landscape painting, the phenomenology of travel, the intrinsic morality of foraging peoples, and the significance of place. Some of these topics are explored in this section. During these years, as he continued to revise and rewrite *Man in the Landscape,* Paul found he could hardly keep up with the nation's rapidly changing ecological consciousness that in just a decade or two had moved from nature perception to the nature of natural resources and, finally, to environmental ethics. After leaving Knox College and accepting a lectureship at Smith College, Paul finally published *Man in the Landscape: A Historic View of the Esthetics of Nature* in 1967. By the early 1970s he had accepted an endowed chair as Avery Professor of Human Ecology and Natural Philosophy in Environmental Studies at Pitzer College, one of the Claremont Schools in California, where he taught and wrote until his retirement in 1994.

꧁꧂

THE FIRST ESSAY IN THIS SECTION, "Digging for Our Roots" (1990), was commissioned by David Hanson and the editors of *Places* to accompany Hanson's color photographs of the Colstrip Mine near Billings, Montana. No project challenged Paul more than this one. From a technical and creative point of view the photographs showing the various colored layers of earth cut through and displayed in surrealistic contortions and hues were beautiful. Yet the images depicted the rape of the earth at its worst. These pictures dragged him back over the bumpy ground of nature aesthetics and challenged him to rethink his original premise. He projected the slides on a screen over and over, studying them pensively, dredging the depth of his feeling and understanding to make sense of them.

"Five Green Thoughts" was published in the *Massachusetts Review* in 1980. The five short sections summarize succinctly the foundations of our ecological perception and provide good homework for environmental types like myself—geographers, landscape architects, historians, nature writers, environmental educators, ecologists, philosophers—who write and think about "landscape." In some of his writing Paul holds a mirror up to us so that we can see our primal being. In this essay, however, he swings open the shutters of a picture window so we can see the terrain.

"The Nature of Tourism" (1957) was published just as Paul made the transition to his new appointment at Knox College and was beginning to travel extensively to sites for research. Sensitive to the increased reaction abroad against "American tourists," he writes partly a self-criticism and partly a justification of the importance of travel. He wrote, the following addendum to this essay shortly before his death in 1996:

> In the forty years that have elapsed since I sought some apology for the symbiosis between certain kinds of travelers and the industry that provides for them, it has become more difficult than ever to defend that dismal alliance. Independent tourists, finding their own rooms and following their own itineraries to roadside attractions and seasonal hotels, could be forgiven with a mote of sympathy for their bland emulation of the old tradition of the pilgrim. But today families go like sheep on tours. Rich, they go with a group by jet to Tibet; poor, with a group on a bus to Disneyland. Urban greed has provided false main streets, and international cupidity will take them to sybaritic island retreats or to walk (with others) among penguins. Only when we remember that the cities themselves are worse and (despite the

standard of living) the conditions of daily life more demanding, can we find a bit of forbearance for the whole desperate, tasteless performance.

"Whatever Happened to Human Ecology?" published in *Bio-Science* (1967), presents a critical, historical review of the path of "human ecology" from its beginning until the late 1960s. In this essay Paul puts a capstone on nature perception and heads in the direction of nature ethics. He attributes primary credit to Marshall McLuhan (with Harley Parker *Through the Vanishing Point: Space in Poetry and Painting* [New York: Harper & Row, 1968]) for this change of heart and mind that came with the "terrible discovery" that nature aesthetics is the mask behind which lies the domination of nature. Rather than drawing us closer to nature, it distances us and detaches us and provides a means for us to step out of the picture.

By the end of the 1970s *The Tender Carnivore and the Sacred Game* and *Thinking Animals* had been published. The beginning of a more mature and critical view of the field of ecology—human ecology in particular—is evident in Paul's next essays. The romantic idealism that had borne him through the 1950s and 1960s was beginning to fade with the graying of his hair. He became increasingly disillusioned with the pillaging of the earth and no longer believed that an understanding of ecology would solve our problems. "The Conflict of Ideology and Ecology" (1977), published in *The Search for Absolute Values in a Changing World,* vol. 1, and "Sociobiology and Value Systems" (1980), published in *The Responsibility of the Academic Community in the Search for Absolute Values,* vol. 2, were presented at conferences on the unity of the sciences. Written almost a quarter of a century ago, "The Conflict of Ideology and Ecology" provides a critique of what Paul called "modern existential cultural relativism." The essay can also be taken as a postmodern view that helps us understand two pervasive global ecological and social problems: mass extinction of species diversity from the earth's ecosystems and the "cleansing" of diverse or opposing views by cultures throughout the world. In "Sociobiology and Value Systems" he defends the unity of knowledge

as he critiques the dichotomies set up by "science" that separate fields
of inquiry and humans from animals and anomalous Others.

For over a decade, from 1968 to 1978, Paul showed the signs of
a change of mind regarding his understanding of the origins of ecological perception in humans. During that time he had been formu-
lating a book he tentatively called "Roots of American Nature
Perception." In 1978, on leave from Pitzer College, he traveled extensively and wrote thousands of words trying to frame the parameters
for this book that would address "the psychohistory of pastoralism as
a framework for Western thought." To do so he went to other parts
of the world where pastoralism and its results were still in evidence.
In the spring he traveled to Gujarat and Rajasthan in India, where
primitive pastoral organizations still flourished. He visited the foot-
hills of the Himalayas in Nepal and Kashmir where animal grazing,
accompanied by clearing and planting, represented the final stages of
what had happened in the Mediterranean two thousand years before.
In the fall he went through all of his notes on the history of ideas
regarding "man/nature relationships in Western thought." Explaining
his activities that year he wrote:

> By the end of October I had completed sixty-five
> 1500-word essays but was dissatisfied with the inte-
> gration. November was spent in Greece, visiting
> formerly forested sites and comparing the deforesta-
> tion-erosion complex with other subarid environ-
> ments. I found the natural setting of Greek thought
> and mythology extremely stimulating. Half-days in
> Greece were spent revising and rewriting.
>
> December was a month of overturn. I put back
> the year's work and reconsidered. The continuity of
> extraordinary destruction then and now seemed
> inconsistent with a simple history of ideas.
>
> By the end of December I had discovered my
> integrating theme. All that was written was put away
> for the future. I completed a new précis of the

"Roots of American Nature Perception." My publisher agreed to the change. At the end, a Commencement.

In 1979, Paul brought his ailing mother, Clara, to live with him until her death the following year. Owing perhaps to the loss he felt for a mother who had unconditionally supported his creativity and development, as well as his remembrance of the nurturance that her passing uncovered, he began reframing his book in a way that would address the rearing and development of children grounded in nature. It wasn't until the spring of 1981, however, when, each morning on his yacht in Morro Bay, with his old dog, Buck, at his side, he finally wrote the book that would become what he considered his most important and most original. *Nature and Madness* was published in 1982.

Paul shows his argumentative side in "Ugly Is Better" (1977) as he takes up once more the theme of plastic trees introduced in "Five Green Thoughts"—but this time with a scathing criticism of our obsessive-compulsive tendencies with cleanliness, tidiness, and consumerism. I, of course, did not know Paul in these younger rambunctious days. By the time I had met him, in 1985, he had traveled thousands of miles, written thousands of words, and ended two marriages with women he had loved and respected. He was a little jaded, quiet and dignified, and somewhat subdued. But he possessed a wry sense of humor that delighted me every day I was in his presence. When he grew restless, we traveled. I learned to pack quickly and lightly since I had to carry my own bags; he was egalitarian to the core. Driving through the countryside, something he loved to do, he was often quiet and pensive, except when he interpreted a bit of geology for my benefit. I took these periods of silence to mean that he was sorting mental files and framing new projects. But the appearance of signs announcing who was keeping America beautiful and clean would often break his concentration and set him in a rage. At times I had to stop him from getting out of the car and tearing down the proclamations that he insisted defaced the countryside much more than discarded rubbish. "Why would anyone want to pick up someone else's garbage? What are we teaching these litterers?" he would rail.

In the concluding essays he emphasizes the importance of place in our lives. "Itinerant Thoughts on Place" presages *Nature and Madness* and explores human development. Whereas in earlier pieces in this volume he emphasizes the role of animals in human development, in "Place and the Child" (1991), an unpublished essay, he turns to the importance of place in childhood grounding and adolescent passage. Both of these pieces show the profound influence of Edith Cobb, author of *The Ecology of Imagination in Childhood* (New York: Columbia University Press, 1972)—a book she had worked on for most of her adult life. Her lifework was published after her death at the age of eighty-two with an introduction written by her good friend Margaret Mead. Edith Cobb had published an essay by the same name almost twenty years earlier, an "abridgment of a longer work in progress," in *Daedalus* 88(3) (1959):537–548. Paul had been very impressed by this essay and asked permission to publish it in an anthology that he and Daniel McKinley edited, *The Subversive Science: Essays Toward an Ecology of Man* (Boston: Houghton Mifflin, 1969). In a brief introduction to the essay, Paul had this to say:

> The relationship of mind to nature is the crucial question. . . . If we deny that mind requires anything in its environment save other minds, we imply that the quality of natural surroundings is not very important and that, indeed, place is expendable. If the uniqueness of place were only a phenomenon, like a passing play of colors, then we could explain aesthetic experience as a matter of attitudes and opinions. If beauty is relative and only skin deep and the beauty of nature actually a kind of recent human invention—a matter of taste—our inherent sense of relationship to it gets lost in a flux of artiness, mode, and fashion. . . . Human sanity requires some less-

than-obvious connections to nature. . . . We have hardly begun to discover what those connections may be.

These words seem written especially for us as we read the following essays on place and carry on our own searches for our fundamental human/nature connections.

FLORENCE R. SHEPARD

Digging for Our Roots

MODERN CULTURE seems about to entertain a turning in its encounter with the hoops that bind humankind with the earth. What wisdom shall guide it?

Both science and religion may have been co-opted and subverted—having become creatures of the exploitation mentality. In a secular society perhaps only art can deal with the problem of evil. But the art that can do this will be one that exorcises its own cankers.

Take, for example, the 500-year-old tradition of the landscape arts. It would seem that therein is a redeeming potential, something that could be enhanced as part of a new colony. But the sources of those arts are the same mathematics that made Lewis Mumford speak of "Galileo's crime." Christopher Hussey makes it clear that the painters took poetry as their source, geometry as their lever, and gave us the picturesque.

The vendors of landscape, being censors and guides of the eye, defined scenery. The painters' evocations of the old poets gave us what would finally become calendar illustrations, doing for the mind via the eye what Muzak would do in the ear. Marshall McLuhan associated this sty in the beholder's eye with the invention of perspective and the picture frame itself—an insider's view through a window in a wall, concealing more than it showed.

We still suffer from that legacy. For out of it came enclave thinking and the sorting of the world into the beautiful and the unbeautiful according to a pastoral imagery that has stood for what is good in nature since the time of Theocritus and the authors of the Psalms. This enclave mentality leads us to preserve nature by partitioning it into parks or wilderness areas. Conversely, this perceptual lock on landscape has conditioned us to surrender willingly all that was unbeautiful to industrial ravagement.

Artists can rhapsodize and paint the debris of the miners and loggers provided they stick with the language of limners' manuals—the basic circle, line, and triangle—or dally with color complements. An educated elite can then admire, in the name of aesthetic abstraction, waste, poison, and death. The rest of us could make do with the picturesque wherever scattered groves remain or sublimity is not upstaged by ski lifts, towers, and gondolas.

But the artist who would break from the stranglehold of scenery and yet avoid the various "traditions" that represent nature as colored retinal images is in a quandary. If he shows the "ugly" reality, "nature" vanishes and the patron turns away. If he opts for abstraction, disavowing mere "subject matter," he risks making evil attractive.

Evil? In *On Photography*, Susan Sontag protests that surrealism is a cruel abuse of real events. She asks whether we are truly free to enjoy old photographs of suffering people—however excellent the photographs may be technically—simply because we have forgotten the people' names and circumstances. These photographs, she wryly remarks, "nourish aesthetic awareness and promote emotional detachment."

Do time and distance make the anguish of people or land-scape acceptable? Should we admire such photographs solely as objects? No, Sontag insists. This "surrealistic enterprise" places the final distance between the observer and the occasion. Talk about form or balance or line if you wish, but there is no escaping the subject.

The abuse of nature in the name of aesthetics has not been lim-ited to art. At about the same time open-pit mining began in earnest at Colstrip, Erhard Rostlund wrote in *Landscape* that clear-cutting forests produces a more beautiful prospect than selective logging or not cutting the trees at all.[1] The Olin Chemical Co. advertised in *Saturday Review:* "If You Think It's Beautiful Now Wait Till We Chop It All Down."[2] Even the paper company was pulling out of the picturesque to help us adjust our connoisseurship to the dissociation inspired by the New Critics. Today, at the Ken-nicott mine in Butte, the signs along the highway say "Historic Site Ahead" and "Technological Marvel—Bring Your Camera." Since humans purposively cut down the land, it follows that the result must be beautiful.

The visitor's astonishment in places like this is an intoxicating distillation of our national power and will to dominate. Like the smoke from the Four Corners power plant, the open-pit mine is an entity that can be seen from the moon. We associate the vertigo we experience at the pit's edge with the exhilaration of our national and industrial success. The camera comes next, for it will turn that havoc into the two-dimensional replica that can awaken an echo of the almost obscene ambiance or, for an audience, provide a concatena-tion of pure visual enjoyment.

The Morality of Hanson's Colstrip, Montana

When David T. Hanson presents us with aesthetically pleasing pic-tures of earth at the Colstrip open-pit mine, Cézanne, Constable, and the Hudson River school are far away. Claude, Virgil, and Theocritus have vanished, too, as though landscape itself had disappeared.

There are various ways to consider photographs such as these. They could be, as Sontag says of the images of Depression-worn families on the road, a reprehensible exploitation of this place purely for visual pleasure.

But I do not think Hanson is trying to flush away an outworn romantic idea and replace it with visual abstractions or symbols of power. Indeed, there is a link between his photographs of these raw earth layers and the heart of objections to the neoclassic logos of rationalism, capitalism, and industrialism.

Somewhere under the husk of romanticism's sentimental excess there lurked a deeper design. Linear thought has so dominated the Western world since the time of Copernicus that its rationality and mechanism have deprived modern culture of the very terms of an alternative. Romanticism was an effort to recapture a lost paradigm: an organic model of creation, a sense of the earth as a living being.

If the land is an organism, what does it mean to cut down a forest or cut open the land? And how do Hanson's photographs escape the moral outrage Sontag expressed against parading distant wounds for casual use in art?

The answers are not simple. First, it must be clear that this new vision or organic sensibility is not simply another ideology, but a new seeing. It addresses the problem of how humans perceive nature and their own human identity.

Although the issue Colstrip raises is similar to that raised by the photographs Sontag criticized, its status is different. Her rage against the pictures of starving people as coffee table amusement was a critique of the whole of Renaissance and literary humanism. The very arrogance and pride of humane morality helped sustain a strong sympathy for downtrodden people—while at the same time widening the gap between the human and the nonhuman. That hubris had never completely wilted before the cold hearts of the makers of the Industrial Revolution and its modern representatives.

Yet the wounds of the earth are similar matter. During the classic phase of environmentalist ideology from about 1964 to 1976 (as distinct from an earlier natural resources conservation), Leo Marx's influential book, *The Machine in the Garden,* examined the ecological movement as a conflict between a "pastoral" ideal and progress. If that were a complete analysis, the whole issue would have become merely another choice of consumerism. But the shift proposed by the ecological movement was more revolutionary. It addressed "mind and nature" and the preconscious assumptions of cultural style, the reawakening of a mythic understanding.

Both the romantics and the ecologists urge us to abandon the Enlightenment's logic and obsession with binary divisions—such as separating the beautiful from the useful or rescuing nature by preserving enclaves of it. (As Leo Marx would put it: understanding the world as a machine.) Once we do that, the problem of surrealist irony may disappear. Art might recover the importance of content, locality, and participation.

As we begin to accept the story of humans as part of the larger story of all life on earth, perhaps we need to search for obvious targets as Mother Teresa does in working among the worst of the sick and dying and as Hanson himself does in photographing.

For the most part, Colstrip is not only a ravaged but an invisible place. So that we do not misunderstand, Hanson has virtually excluded people from his pictures. Like Cézanne before Mont Ste. Victoire, we are drawn to form and color, to the brink of that alienated mood Sontag chastised. Hanson starts with the degree of dislocation that an ambivalent culture has found aesthetically acceptable.

We seem at first invited to scrutinize a juxtaposition of mining and the human environment it creates. But do they have this apartness? The lack of people in the photographs prevents a certain kind of distancing: The absence of other "selves" makes our involvement as viewers that much more undeniable, just as the absence of the self in the animal dreams of young children makes the dreamer's presence more vivid.

What we see has neither the emblems of romantic technophilia nor romantic grandeur; it is not even a landscape in its customary sense. We are pulled up short by the estrangement caused by the objectification and abstract detachment of these photographs. Just who is the wounded and the wounder?

Compare these photographs to Alexander Hogue's painting of

the raw torso of rural Oklahoma during the Dust Bowl era of Mother Earth Laid Bare. It was a primitive effort that must have brought abusive chuckles from the avant-garde enemies of subject matter and content. Hogue's evocation of prudery in redressing a land denuded by the plow was quaint, but one knows he had rape in mind.

Something more horrible, an act we have committed upon ourselves, is at hand in this evisceration. We have begun to escape the metaphor of the earth organism as poetic convenience and to recover its meaning in homology, in a common ground differing only in expression.

Healing the division of the world into what is pictorially aesthetic and what is not begins with the act of attention. Our eyes, educated in Anglo-Americanized Italianate escapist pastorality, still glide quickly past the "ugly." There are plenty of geographer-traveler-writers who tell how "interesting" it all is, providing relentlessly humane description. Such endless fascination with ourselves and our works also educates the eye, but its perception is that of linear analysis. Ostensibly value-free and demythologized, it actually is a perverse enchantment, its mythic core the body of stories of domination that define the West.

Digging for Our Roots

Hanson's *Colstrip, Montana* series is a worst-case scenario that alters our awareness and casts attention to a violence that shakes our complacency. Nonhuman life—animals and plants—is difficult to perceive anew because its identity is clouded by romantic humane individualizing and "Disneyfication." Instead, the new reality emerges raw, elemental renewal, sensitivity to air, water, and earth.

There is a paradox in this kind of backing off from life: a distancing to get close. It is a precursor of a different consciousness: A world of beings, in infinite and mysterious acts of connection, created us from the earth, itself a being.

There is no dichotomy between the mineral and living. Such is the wisdom of all stories of beginnings. Peabody Coal made no acknowledgment of this when it removed the coal from the earth, but the geology of such a cut speaks of the earth's anatomy. The Rosebud Formation, a 24-foot-thick seam of coal—the remains of an incalculable host of plants—is uncovered 100 feet below the surface. That 100 feet, the "overburden," is misnamed. The real overburden is 3,000 years of human estrangement from nature, nurtured by bizarre fantasies of human identity. "Rosebud" could not be a more ironic name for the whole black mass of mummified plants that once swayed in the surge of a tidewater sea covering the center of the continent.

More than any other understanding, it offers the perspective of time. At this undersurface, 60,000 times 1,000 years ago, a swamp hid the last dinosaur bones, now visible again. Below this cadaverous mass, tan and gray sandstones and mudstones that give so much of the color to Hanson's photographs were accumulating when something monstrous struck the planet.

Hidden in these gray rocks is a thin layer of iridium dust: the remains of the impact that changed the history of life. Dust, blasted into the sky, surrounded the earth for months. In this twilight the plants died, dragging their animal dependents into oblivion with them. A minor scavenger form survived, giving rise to the birds. And within the ecological vacuum, our furry quadruped ancestors rummaged in the dark, eking out their survival on the bodies of worms and bugs.

When the Rosebud Formation was still a marsh of living sphagnum the continents had not drifted apart and together were an island in a world ocean. On Pacific atolls, Margaret Mead once observed, islanders know that an island is the symbol of limits. On Pacific atolls, islanders know about bounds and have a sense of scale in their affairs. Seen from space, Earth is an island. That view is surreal, like Hanson's photographs, and less a staircase to new frontiers or a means of dismissing the earth than an effort at insight, a prelude to recovery, a reminder of our limits.

Like space travel, the surrealist vision is dangerous ground. For, as Sontag says, it is a callous denial of the passion of lives lived and a celebration of forms, postures, and compositions. The visual allure of photographs like Hanson's is addictive, says Sontag, and can turn us into image junkies lusting after "an amorous relation, which is based on how something looks" instead of what it means.

But risk can have its rewards. If we can avoid translating Colstrip's awesome forms into admired geometry we may see beyond either forms or pictures. Hanson's lens is sharper than bulldozer blades and giant shovels, for it enters our heads to open seams, to look for grounding. (Where has the acid rain from the 133 million tons of coal removed from Colstrip fallen? How many shabby ex-mining towns are there, and what living death haunts them?)

In inviting us to look in order to perceive truly, Hanson traffics in the colors of poisonous effluents, like a shaman curing with the glands of toads. He asks us boldly to exercise a kind of hue-delight as a means rather than an end. At first we are reminded of the mineral brilliance of Roman Vishniac's microphotographs of translucent slices of minerals at the other end of the size scale. But the conjunction of detailed captions and the series of images links Hanson's work, not with painting, but with the narrative arts. It is Our Story: a recovery of social and ecological truth too long repressed by the industrial-technical era. We are awakened to patterns collecting us in a violent tale of time and place.

Notes

1. Erhard Rostlund, "The Changing Forest Landscape," *Landscape* 4(2) (1955):30–35.
2. *Saturday Review,* April 8, 1967, Olin Chemicals ad on page 1.

Five Green Thoughts

The Sheep as Custodian

The idea of an Ovine sentry came to me as I stood on a large ring of earth in Southern England, trying to imagine the medieval castle keep of which it was the ruined foundation. Sheep grazed quietly in the center, and it was *their* turf that protected the relicts buried beneath from the erosion of the centuries. Their droppings fertilized the green shroud. Like a company of mild wardens they occupied the space, giving it an air of permanence, of keeping.

But that air, like the static feel of a museum, is an illusion, for such places are the sites of old havoc, piles of the dismembered and disjointed. The unspoiled–looking site had long since been sliced through and the old mounds built up again.

Why must one dig up the past? How and by what agency do all those fragments get buried in the first place? In the Western world in particular our sense of history is intimately related to exhumation. Of course, graves are a part of the explanation, and the walls of villages and cities from England to Teheran were built on top of the ruins of predecessors. Fallen ceilings became new floors in a hundred cities across the Mediterranean world. Our roots are down there and we—represented by archaeologists—are like moles, otherwise blind to our origins, tunneling and nibbling at them.

Any archaeologist will tell you that relatively little of the remains of seven thousand years of town and city life actually got buried. Most of the products and belongings of vanished peoples simply disappeared, not buried, just kicked around, worked on by weather and time, carried away, broken up, dissolved, disintegrated. Yet the fragments of old pottery, hardware, and masonry were not randomly buried. Much depended on the local drainage: the city's place in the watershed.

The mortality of those ancient cities was itself related to the impact of people and their animals on the terrain. Throughout the "cradle lands" of civilization, the agriculture upon which the city depended developed a tangled network of canals, reservoirs, gates, and channels presided over by engineers, slaves, and bureaucrats.

The elaboration of the state was dependent on making land usable, and it in turn supported greater populations. As irrigatable land ran out, the tensions between neighboring tribes and chieftains escalated; and as human numbers overflowed the centers, the surplus went higher up the watershed. There they cut timber and grazed their animals and sent down wood, skins, wool, meat—and the soil.

Upstream denudation, valley floor saturation, and conflict: The debacle came when some war, famine, or epidemic combined with the weather disrupted the hydraulic works.

The pretty sheep in grassy places represent the whole tribe of "hoofed locusts" who have toppled so much of the highlands down upon doomed societies. They (the sheep, goats, asses, horses, cows, mules, yaks, camels) administered the internment of eleven successive

Mesopotamian empires. They were the barbers or scalpers of the Judean highlands, the Peloponnesian and the Syrian upper slopes, where their dexterous toothwork and footwork buried cities. King Solomon paved the way 3,000 years ago when he sent 80,000 axmen and 70,000 haulers to take cedar and cypress from Phoenicia. Two thousand square miles of forest was thereafter reduced to 4 tiny groves today, the largest of which has about 400 trees. The livestock followed on the loggers' heels and sent forest humus into the sea— or into the valleys where it simply buried the pillaged and burned remains of whatever army happened to have swept through.

In such a series of debacles are many generations. The magnificent forests of the Mediterranean rim and islands were progressively demolished and their seedlings and root-shoots chewed and trampled by livestock. The relicts of that vegetation—the maquis (myrtle, box, oak, olive, and oleander shrubbery) and the even more degraded garigue (heath, juniper, pistachio, viburnum)—cling to raw earth between the rocks, giving an appearance of timeless austerity. The blanket of soil, once the sponge for a million springs, vanished so long ago that even the educated traveler, who finds the region so picturesque, considers the rough, raw land as "natural."

Still the classicist writes of the "puzzle" of the demise of Minoan civilization. The historian calculates the political and military factors that destroyed Jerash, now a village of 3,000 that was once a city of 250,000. The cities of Mesopotamia faded before bloody invaders and "natural" catastrophes.

It is possible that in some places this cycle of buildup and collapse was run through relatively quickly, while in others there were not periods of stability when the downstream mainworks and the upstream plundering were so intense and a stable ecological relationship prevailed. Such places were, so to speak, in the process of not being buried. Having left us less to dig up, they are less well known and, being unknown, are omitted from our official history.

So we come to the inverse relation of land use and history: The worse the land practices, the more surely the "culture" was buried. No wonder Western consciousness is an overheated drama of God's vengeance and catastrophe, preoccupation with sacrifice, portents and omens of punishment by a heavy-handed Jehovah. Like the dinosaurs, who are known mainly for their vanishing, the ancestors we know best, and from whom we take our style, are those who seem to have lived mainly to call down calamity upon themselves.

The whole thing seen from the standpoint of the goat or sheep might seem utterly reasonable. First the centuries of teeth and hooves, slowly cutting loose the sides of the mountains, and finally their own silent presence in the lowlands, like woolly old museum docents, inane munchers, watchful angels over all that stuff filed away in the basement.

Why the Greeks Had No Landscapes

The lack of landscape painting does not mean that the ancient Greeks lived suspended in the sky with Apollo or spent all their time at sea with Poseidon. They, like ourselves, had their feet on the ground most of the time. It refers to a lack of description, of setting or scenery, in literature, poetry, and painting. The Romans, for example, had whole walls covered with mosaic views and a tradition of pastoral poetry with images of the country retreat.

This seeming lack of attention to the terrain, or accounts of plants and animals as part of narrative or myth, led early scholars at one time to conclude that the Greeks were either not interested in nature or had no talent for exploring it. "The literary genius of Greece," writes Walter Greg, "showed little aptitude for landscape, and seldom treated inanimate nature except as a background for human action and emotion." Together with the ancient relics of the sculpted human figure, pottery painting that depicted figures but not place, and "classic" architecture, this absence of surroundings was seen as part of Greek narcissism, and Greek interest in the relations between humans and God or between human and human.

In Homeric epics there is no "scene," no purely external description, no nature at all in our mode of portraits of place, picturesque settings, or charming atmosphere. The nonhuman context is not an aesthetic container of the action. Homer does not depict events against a background, as in the barren sets of a play by Samuel Beckett, or conjured scenery, as in Virgilian poetry. For Ulysses, objects are never only things. Hence things are not simply acted upon; they participate, standing out brightly for the sake of the relations among them which are messages from mind to mind.

For the old Greeks, nature was not "treated" by the artist but was for all consciousness part of the animation. Creatures, plants, rocks did not surrender their otherness to scenery. They abided in their own right and moved in relation to the psyche, converging without simile, a true kindred reality, never merely symbolic. All terrain was an extension of what men experienced mentally and metaphysically, occupying places of existence but not different parts of a dual reality.

Paolo Vivante, in *The Homeric Imagination,* says that we must imagine a consciousness in which there are no literary associations, but in which all events, human and otherwise, derive from common vital principles. The human feelings are expressions or extensions of more profound commotions. Greek verbs combine human and natural action, which have both a psychic and physical meaning. The word for "melt" refers at once to a state of snow and human tears. Things are not "like" other things but share their process.

Although there is little reference to the features of particular places in Homer, each place is profoundly unique. The sanctity of the earth is locally signified by the *memos:* a place often defined by cleft horizon lines in which a goddess is immanent. Divinity resides, characterizes it; mind and feeling have worked upon it. Shrines are possessed by deities, not built for them or dedicated to them. Because of the universality of events, human action is never separate from its other forms. Thus intense human action, says Vivante, may stop while they attend to the glitter of bronze or the resounding sea, which are part of their action. In this the significance of one thing is heightened by the mutually responsive qualities in another.

Our modern perspective of space and time in which human action is located and described may also have some of its roots in later Greek thought—particularly in Plato, in Euclid, and in certain aspects of Greek theater with its scenes and its parabasis, in which the chorus removes its disguises to comment on the action. It is this detached observer for whom the pictorial view become possible.

Socrates scorned the old oral traditions. By the time of Periclean Athens the Greeks may have been philosophically capable of conceiving landscapes. The Roman frescoes of the first century are Greek thought in pictures; painted portraits involve a distancing and eventually a setting. In Flemish painting of the fourteenth century there appears beyond the faces of saints fragments of landscape, the harbingers of a new subject matter. The revolution in visual thought, keyed to mathematics, was first incorporated in panels painted by an architect, Filippo Brunelleschi. This perspective of distance was discussed theoretically by Alberti Leon Batista in 1435 and came to be called Euclidean, distance-point representation in which space was "mathematically homogeneous" and ruled by the "laws" of convergence. From this, various other abstract rules developed for describing the "unity" of pictures—and hence of nature. Marshall McLuhan has been at pains to point out how the application of mathematics to space organized the visual world into perspective, isolating the seer from the seen and creating secular space and human alienation from nature. "Civilization," he writes, "is founded upon the isolation and dominion of society by the visual sense." This visual sense is associated with a fragmented human identity. Outside the picture, viewers become onlookers, spectators. They become connected to the picture itself, not to what is shown, a part of a work of art. Life as a work of art was born. Its relationship to nature is a *stasis:* a presentation of selected fragments of visual experience. Like literature, and through literature, the "scenes from a life" are connected by a story.

Panofsky has said that the painting—unlike its medieval predecessors—became a kind of window. The modern reader may find this idea confusing, as we now think of windows as *connecting* us to nature. What it connects us to, however, is landscape: The window makes real the wall as a separation, its fenestration a calculated visual portrayal of certain external "symbolic" objects. An interior with no windows is a whole canvas. The unity that is achieved is of physical space, while what is fractured is the continuity in time without which there is no life. The optic culture tries to repair this by an association of the scene with a literary idea or story, but the repair is purely intellectual. It is not lived. The multisensuousness, especially the auditory quality of the Homeric world, did continue in Medieval Europe as *participation,* in which speech and the word continued to be a shared principle in the whole of nature.

For most of Western history, two different approaches to experience existed. One was evident in Homeric and primitive thought. The other, apparent in art in Athens as early as the fifth century and inseparable from Euclidean space, increased with the phonetic literacy and was essential to the idea of nature as distinct from humans. A contrived picture of the world is a landscape, which in our time we have come to think of as synonymous with "nature."

Virgil in Westchester County

The appeal of suburban (or exurban) life is that it provides a bucolic setting without the concomitant monotony and trash of agriculture. Its popularity implies that we share a pervasive, almost compulsive ideal—not an ideal hammered out of personal reflection or social dialogue, but one that the culture imparts. It is an adjustable ideal. Basically it is composed of a large yard (from *jardin,* or garden), separating the house from the street, and extensive unbroken but undulating meadow with scattered trees that sets the house off and yet blends it gracefully with the surrounding countryside. Subordinate buildings, pools, stream, rockwork, paths are incidental to it. Its original theme was the association of hoofed animals and dogs in a tight symbiosis with people. What it does not include in its most perfect expression is naked wire or steel fences, cultivated ground, standing crops, farm machinery, fuel tanks, storage bins, loading docks, junk or garbage piles, woodpiles, manure heaps, old vehicles, wheelbarrows, wagons, a cordon of old barns, electric wires, washtubs and clotheslines, pigs, or any other paraphernalia of the working farm.

Yet the suburban concept is not derived from a tidied-up farm. It is not a sort of Puritan or Dutch housewife or modern sanitized version. It has wholly other sources that come not directly from agriculture but via an urban dream and its aesthetic consequences interposed between distant rural antecedents and middle-class taste. Without being aware of it, even the educated inhabitant knows little or nothing of the convoluted history of The Pastoral.

Modern writers sometimes use "pastoral" to mean anything outside the city, from farmland to wilderness. Actually they know better, for the pastoral is a literary genre. This slipshod use of the term is uttered deliberately to minimize the variety of nonurban landscape by those whose theme is the contrast between the city and the country. The genre has a long history. Its lineage includes Theocritus, certain Romans, and a number of poets and dramatists: Boethius, Jean de Meung, Boccaccio, Sannazaro, Spenser, Sidney, and Milton. Its best-known exponent is probably Virgil, whose name, for anyone with a smattering of literature, conjures just such idyllic scenes of shepherds and their animals that the layout of the modern country house is intended to imitate.

What is one to say today of a two-thousand-year-old body of poetry that alone created the images of the idyllic and bucolic that still fill our heads? How are we to deal with the magisterial pronouncement of Johan Huizinga that "however artificial it might be, pastoral fancy still tended to bring the loving soul into touch with nature and its beauties. The pastoral genre was a school where a keener perception and a stronger affection towards nature were learned." But what "loving souls" does he mean—shepherds themselves, urban people, naturalists, or just lovers? And what "nature" does he refer to—the denuded slopes of a half-million square miles of Mediterranean watersheds?

Theocritus knew the real harshness of the Greek terrain so well that he placed his poems in Sicily. The Roman, Virgil, no less aware of the gap between the Campagna and "nature," set his version in Arcady. Only if you put it far enough away could the illusion of goats scrounging sprouts among the rocks be accepted as "nature." Huizinga must mean educated loving souls whose "keener perception" of classical lands was that enhancement of landscape through Virgil's spectacles and those of the great "naturalists": Petrarch, Boccaccio, and the rest who used the pastoral for tales of social intrigue, satires of prominent politicians of the day, nostalgia, allegory, escape, a model of the Christian paradise, or other metaphysical imagery.

The main connection between that ancestral body of defunct literature and modern expression is by way of certain landscape painting and, in turn, landscape architecture. This series of emulations, each removing the modern reality farther from its model in ancient Sicily (Theocritus' boyhood home) has been written upon at length. It can be quickly reviewed in Elizabeth Mainwaring's work and its American extensions understood in the influence of nineteenth-century architects such as Andrew Jackson Downing and Frederick Law Olmstead. The point is that ancient reverie about pastures made its way through the arts steadily and perniciously, emerging finally as the stereotype for suburban dwellings and rural beauty.

It is ironic that the term "pastoral" has come to symbolize the whole of nonurban environment, for its antecedent has no equal for ecological destitution. Looking at the epitome of its type—the country house with its graceful horses and purebred dogs—it is difficult to believe that this pattern and style of the occupation of habitat is the product of a great ecological lie.

As for the social metaphor made from the relationship of humans and sheep, Aldous Huxley makes the rueful observation: "We go on talking sentimentally about the shepherd of his people, about pastors and their flocks, about stray lambs and a Good Shepherd. We never pause to reflect that a shepherd is not in business for his health, still less for the health of his sheep. If he takes good care of his animals it is in order that he may rob them of their wool and milk, castrate their male offspring, and finally cut their throats and convert them into mutton." How is it that an economic activity which, as much as any single factor, is one of the most destructive forces in the world could be so beautiful? Anyone who wishes to look at the evidence can satisfy himself that grazing and browsing animals have been agents of the collapse of Mediterranean empires, nations, and city-states and the principal means of the impoverishment of equatorial lands more decisively than all of the wars. On any slope at all the goat is simply the nemesis of the land that he sends rapidly off to fill deltas, estuaries, and lower river valleys around the sea.

The lie is that the goat and its fellow grazers and browsers and the denuded and eroded lands are an enchantment. It might be imagined that the ancient Greeks and Romans, the Renaissance Spanish, or the modern Moroccans and Syrians and American Westerners are blind to it, but the record does not show it. On the contrary, there is written evidence in every age that some people have been totally blind to the effects of overgrazing, but there has always been evidence, too, that some people could see what was happening. It was not a question of ignorance.

Rather, I think, the pastoral fraud is based on a selective vision. Whatever the motivations for perpetuating the geological calamity of grazing, that economy over the centuries evoked artistic expressions. The most deceptive feature of such land systems is their stability. Economies create their own justifying demons. Indeed, pastoral art has often been criticized as "static": Things go on for centuries without apparent change. Large ecosystems do not vanish, but they do change, often imperceptibly. Their productivity and composition diminish to a low equilibrium. So there is built into this complex of flocks, fields, and herdsmen an image of endurance: the abiding earth. The city has often been sacked and burned, redesigned and rebuilt, and destroyed again. Even the farm was altered—or disappeared as its fertility was exhausted. Against these riptides of political turmoil and upheaval, the merchant and bureaucrat found unfailing consolation in the peace of meadow and sylvan glen, flocks and murmuring brooks, piping shepherds, sunshine and birds.

That the shepherd was usually brutal, hostile, stinking, and stupid, that the order of magnitude of destruction was more like that of glaciers and climatic change than of battles and plagues, that the potential of the land for human well-being was degraded under asses, sheep, cattle, horses, and goats—these were not what the harassed bureaucrat, military chief, or tradesman wanted to see. What they did see was physical evidence of Elysium: a land of leisure where drudgery was unnecessary and bribery and conniving unknown or innocent by comparison. They needed respite from the stench of cities without sewers, from rats, epidemics, assassins, noise, and the treadmills of survival. Compared to all that, the dung of cattle was sweet and the vacancy behind the shepherd's eyes a relief. Things could have been seen this way especially when the magic of art could make it so.

Theocritus, roasting in Alexandria, was dreaming of his boyhood. Virgil is not known ever to have held a hoe in his hand; he was a gentleman farmer. Spenser and Sanazarro were allegorists and Pope a dandy. Bellini and Claude, the landscape painters, were not doing portraits of place but imagined scenes from Greek Arcady or the biblical. The landscape architects like William Kent, Capability Brown, and Humphrey Repton, using plants instead of paint, mimicked scenes already twice-removed from the physical world, first by the poets, then by the "classical" landscape painters.

Theocritus had been interested in the play of joy and sorrow in life, while Virgil is said to have "discovered the evening," that is, to have associated melancholy with certain terrain effects and with dusk or time-gone-by. Thus the past instead of the distance was associated with the elegiac sentiment: The twilight reminds us solemnly of the idyllic world of long ago.

The emphasis on images and vision tells us that what was sought was a picture of that eternal prospect in the mind's eye: a world made simple and endlessly comfortable. Since the world, naturally and socially, is not in fact simple, what can one do to realize it? The goat, under our guidance, has solved it for us. It amounts to a kind of lobotomy on the land, done not with scalpel but with teeth and hooves. There is the victim, placidly like an ex-patient in a threadbare green robe, no longer full of primeval thunder and night creatures, all sunny and sweet, or at least mild, and, as they say, "spacious."

Lawn of the Giants

Aldous Huxley once observed that Wordsworth would not have felt the same about nature had he spent some time in the tropics. Its density and fecundity would have swamped his sensibility. Enlarging on

Huxley's thesis, it follows that any culture tends to form standards of judgment about the environment from its own habitat. When individuals encounter vegetation and terrain forms radically unfamiliar, they often feel uncomfortable. One reaction is swiftly and unconsciously to ransack the clues from their heritage: sizes, forms, images, utility, working assumptions about how their ideal standard types would have to be altered to make something like what they see.

Among the habitats for which Anglo-Saxons were unprepared, besides the jungle there was the steppe. Nowhere in temperate-climate northwestern Europe is there sufficient natural grassland to provide an adequate cultural perspective. Not since the Hun horsemen overran much of the continent in the fifth century had such memories been fresh. Moors there were, but their vegetation was typically a heath association, mostly shrubs and herbaceous plants. Grasslands would, by that scanning computer in the European mind, be assumed to be the result of human action and to have replaced the natural forest. Whether that was felt to be a loss or a gain depended much on individual perspective, as we shall see.

The historical test case was the grasslands of North America, particularly the tallgrass prairies of the midwestern states. As emigrants expanded into this new frontier west of Ohio in the early part of the nineteenth century they found, first on the uplands, openings in the forest, then intermixtures of scattered trees and grasses, and finally great stretches of prairie, sometimes miles in extent, with scattered groves of trees and shrubs along the streams. What could one say about such a place? How to explain it? Would it be a good place to live? How did it strike the itinerant's sense of beauty and utility? What did God have in mind here and what could humans do here?

Much of the earliest Caucasian experience in the prairies was by the French Jesuits, who gave them their name. But the Jesuits in the eighteenth century were merely traversing a foreign, heathen land. Although they wrote of the soil, forests, grasses, and fires, their main concern was with souls and the comments are brief: "beautiful," "verdant," "boundless," "frightening," even "pleasure grounds." But in general the Jesuits were not looking for a home or trying to make sense of a place. It was the later settlers and eastern visitors who puzzled out appropriate descriptions.

The pleasure-garden theme was certainly one of them. With any education at all, the traveler saw "a magnificence of park-scenery, complete from the hand of Nature, and unrivaled by the same sort of scenery by European art." Even "enchanted ground," of all things, grasslands "gemmed with wildflowers of every hue, the stately forest and valleys interspersed with shady groves—the wild and bounding deer in great numbers" where the eye "wanders from grove to grove, charmed and refreshed by an endless variety of scenic beauty." And in the rocky bluffs "along the banks of the Upper Mississippi there stretch for hundreds of miles the ruined facades of stately castles and magnificent temples, built by Nature's hand."

One was, I suppose, at once Adam and God's Protestant elite. "I have no morbid sensibility," said William R. Smith, "on the subject of taking possession of a land which was worthless in the hands, and under the dominion of roving savages . . . a country so recently rescued by the enterprise and valor of our hardy pioneers from the wandering Indian, whose only occupation was to hunt deer and spear fish, although dwelling in a Western Eden."

Then came the cultivation of Eden. The moldboard plow, which could handle the tough prairie grasses, was invented. Some were not so sure of the fertility of a soil that did not support trees. "Oak barrens" they called the savannas between prairie and forest. It was clear to others, however, that the annual prairie fires caused the lack of trees. Along the prairie borders the first farmers planted themselves close to the woods, somewhat off the fire's path and the forest edge that hovered, to their mind, like a protective mother. "As I rode leisurely along upon the prairie edge," wrote an itinerant, "I passed many noble farms, with their log-cabins crouched in a corner beneath the forest."

For those with an eye for the sublime spectacle, the great prairie fires were indeed exciting. So were the storm, and the wind that made the grass look like sea waves. William Cullen Bryant put it all in his poem "The Prairies" and more—beyond its sheer sensory impact there was the mystery of its presence. Taking his cue from the signs of a shaping intelligence, the past he created was that of pastoral civilization gone to ruin, its builders vanished, its livestock degenerated into buffalo and "prairie chickens." The land was now to be rescued from its barbarity.

Despite the extraordinary effect of the prairie "on the mind" it was doomed. The ideal of the garden granary of the world would sweep away any persisting notions of pleasure gardens as fast as hedgerows could be planted or fences built. Robert Ridgway wrote, of a return to the Fox Prairie near Olney, Illinois, twelve years after an earlier trip, that "the change which had taken place was almost beyond belief. Instead of an absolutely open prairie some six miles broad by ten in extreme length, covered with its original characteristic vegetation, there remained only 160 acres not under fence . . . the entire area was covered by thriving farms. . . . We searched in vain for the characteristic prairie birds." And he noted that the same was true throughout the state. "The buffalo has entirely left us," observed a writer in the *Illinois Monthly Magazine* in 1830. So had the elk. The beaver, otter, and bear were scarce. The badger, like the smaller birds, would survive in fencerows.

So the hypothesis with which we began must be modified. What do invaders in an alien habitat do about its strangeness? They concoct explanations for it, and may make it an imaginative and aesthetic experience—but only if they have no practical interest in it. In the latter case, they set about converting it to look like—to be like—the thing they know. The tallgrass prairie of Illinois, in which a man was said to have to stand on his saddle to find his cattle, was exterminated as a natural community within a lifetime of the beginnings of the settlement. Its extensions in Indiana, Wisconsin, Iowa, and Missouri met a similar fate.

Some tiny bits of prairie have been recreated by colleges and universities from its scattered species. But the opportunity to smell, hear, and see it through the seasons, especially to walk alone in a prairie that extends to the horizon, is gone. The soil the prairie created is too valuable to leave it to nature.

Plastic Trees and Plastic Minds

Seen too close in time, it is hard to know when an era ends. It was indeed becoming clear by 1980 that a remarkable period had passed. *Science,* the weekly journal, carried a full-scale piece called "What's Wrong with Plastic Trees?" The economic cowboys, harshly criticized by environmentalists during the previous fifteen years, were back in the saddle again.

The plastic tree article was about as sensational as that publication gets. But despite its trendy intellectual chic, the viewpoint presented was not one that simply blew in on the winds of change. The argument that the artificial and man-made were as good or even preferable to the natural touched a theme that could easily be traced to the ancient Greeks or beyond—a theme that had surfaced strongly in Western thought ever since Francis Bacon made it clear that the purpose of nature was to serve humans. But the splash with which it hit the fans of public dialogue signified not only some shift in rhythm but a pressure that had been building since about 1965. The barons of technology, the main targets of that crescendo of environmental voices, had been caught off guard. Intellectually and academically they were ill-prepared. It took a little time for them to build a better rationale than "growth is good." By the late 1970s their festering had matured and they had their own stable of scholars and adjusted PR. The era of ecological conscience as a popular movement was to have on its tombstone an avalanche of advertising full of logical, humane, aesthetic, reasonable, economic, and philosophical tirades. If Advertising is nothing more than the spindrift by which one judges the temper of the sea, beneath it were all sorts of theses, doctrines, programs, and all the dialectics pertinent thereto. A great overkill of justification and vindication of The System had begun that seems to go on forever—at least to the graying warriors of the "ecology movement."

One sensed the plastic trees argument was a bit tongue in cheek, but the assertion is interesting at the heart—or the root—of the claim that human works must be consonant with ecosystems and build themselves into systems that are millions of years old. Much of technomania defends itself on the grounds that it loves nature more than anybody. But the old Faustian thought is that the whole kit and caboodle of nature is out of date and will eventually have to go. With so many garden-clubbers, backpackers, and fishermen about, that cannot be said too brazenly, for the engineers of the new world know that otherwise they would soon find themselves knee-deep in restrictive legislation. The plastic tree is one small step toward making mankind the only living thing in the new world order.

It is a clever wedge. Fourteenth-century Arabs were extremely fond of mechanical birds: delicately made gadgets that sang and moved. No one can imagine Islamic science half a millennium ago as forging ahead in the area of metal avianoids as whimsy; it was more likely a response to some felt need. My guess is that three or four thousand years of land abuse, combined with a harsh, dry climate, had so decimated most natural habitats that the birds had vanished. There is abundant evidence of the love of luxurious gardens and parks stocked with animals of all kinds by those who could afford them. Where that was not possible a finely done likeness was perhaps the logical alternative. There is certainly no reason to suppose that Islamic taste would find the idea any more objectionable than that of Christians. Perhaps it makes the lack of things natural easier to take.

If I were an all-powerful king convinced that the pests eating my well-regimented crops or parasitizing my highly bred animals were transmitted by wild plants and animals, that my concrete pavements and asphalt surfaces were cracking from the barbaric forces of living roots, that random photosynthesis was intolerably inefficient and other creatures were still getting half of the energy produced, that the weather was too accidental for human good and needed to be regulated, that the prodigal wafting of seeds and wanderings of animals was inimical to a well-planned society—if I were a benign tyrant I would put a stop to it all. But not in such a way that would alarm my subjects (with their primate bodies and Pleistocene minds).

I would begin by declaring that the world was beautiful—indeed, that its beauty was its most valuable attribute. I would lament that it was difficult and costly to keep things growing in the many perimeter and decorative plantings and little parks that accompanied enlightened progress. Direct impact by trucks, careless tromping by many feet, buildings below ground and roofs above, cables and pipelines, the lack of pollinating insects for flowers, the disappearance of symbiotic fungi and other necessary partners, the very air itself— all seemed to "do in" those things we all love, nay, those necessary aesthetic elements in a highly developed world. However important we find film substitutes and science backdrops, garden wallpaper and organic motifs in finishes and surfaces of all kinds, we must keep some truly beautiful, real landscapes. That I would insist on.

"Landscape" is the pivotal word. It is not a synonym for habitat or place, certainly not for natural community or ecosystem. A landscape is a representation of a certain kind of visual experience. The word was first used to mean a painting. The earliest of these "landskip" paintings (in the West) were imaginary scenes inspired by biblical and classical themes—that is, literary images. "Scenery" itself refers originally to a Greek term for stage props. Thus landscape is a way of tangibly representing a dramatic moment by using terrain and sky and plants as stage props, eventually as actors. Gardens made to be seen, or which are designed around viewpoints, are another tangible expression of the idea.

H. V. S. Ogden and his wife, Margaret Ogden, have documented the historical sequence by which a body of criticism and language of connoisseurship developed first around the imaginary or ideal scenes and were then applied to actual places by Dutch itinerant limners doing portraits of country houses for English aristocrats in the seventeenth century. A place that looked good as a picture was regarded as "picturesque," and Christopher Hussey has shown us how such pictures trained the eye or became a veritable language for an aesthetic of nature. Others have unraveled the profound interconnection between paintings of idealized scenes and the rise of a style of gardening in eighteenth-century England more completely visual than anything previously done in Europe. That pictures were the standard by which raw nature was to be judged and its aesthetic value established is a trap escaped only by a few (of which Henry Thoreau was one). Bernard Smith has traced the steps by which a whole section of the planet—the South Pacific—was perceived by Europeans in terms of pictures.

My program as a kindly monarch would be to dwell on the idea of landscape, to use the terminology of analysis and description it shares with the arts, to talk about aesthetic and spiritual values as identical, and to refer to art, artists, inspiration, and creativity as the supreme human qualities, the great achievements of civilization. When the natural came up, I would assume that it referred to that idyllic relationship between humans and nature exemplified by pastoral poetry. Even ferocious beasts, as seen in the paintings by Delacroix and Henri Rousseau, have a "place" in nature, that is, in parks created for their "preservation." Everything depends on the manner in which nature is "treated," to use the language of art criticism. We all want harmonious, dynamic, balanced, integrated relationships to nature. (Isn't that what ecology says too?)

For a few years—and it may have some permanent effects—a wave of doubt swept across the modern mind that nature and art had the same purpose. There was a sort of rebellion by some scruffy types who agreed with Thoreau that they did not find the play of light and shade on a dead horse in a meat stall to be as satisfying as sunshine on a live horse in a field. Those "romantics" were barred from the technocratic centers of power as Thoreau's century ended. Today we can imagine the hee-haws with which Thoreau would have greeted the substitution of astroturf for meadow. And yet it makes perfect sense so long as we regard the purpose of the world as a setting for the human romance—or tragedy—and order to be a matter of form.

The reverse romanticism of plastic trees is not one of birds and flowers as moral lessons and kindred beings, but a vicious fantasy that justifies the extirpation of nonhuman life. No one doubts that we will have our plastic trees and that they will do even better than seeming like the real thing, for they will be horribly real.

The Nature
of Tourism

Among the most sensitive observers of man and nature are the seasonal ranger-naturalists of the National Park Service. Put upon by waves of holiday flotsam surging into the parks, the rangers continue to translate something of the cosmos with an astonishing display of altruism. With such firm and gentle dispensation do they work that the acid cup rarely overflows. When it does, usually at the end of a day filled with foolish questions and futile explanations, the opprobrium appears which is otherwise checked with magnificent patience. A ranger of my acquaintance was handed a bouquet almost in the shadow of a no-picking sign. Asked to identify the flowers, this green-suited, erstwhile schoolteacher flung them down and twisted them into pulpy fragments under his heel.

The implication of such behavior is not only that the tourists are mistaken, but that they are boobs. Sympathetic colleagues and cynical tourist industries would undoubtedly agree. Historically there is some justification for this conclusion. Before travel was so corrupted, the original tourists to Rome were either scholars or young aristocrats, or both together, one as tutor. The young sixteenth-century gentleman was sent for the "benefit of wit, for the commodity of his studies, and dexterity of his life." This design and its humanistic, didactic formula deteriorated in the following century. Men and women of the aristocratic and mercantile classes usurped the Tour to Rome to absorb culture and get a change of climate. The Grand Tour became fashionable. Americans of the upper social strata patronized the fashion even though it cost them an ocean crossing.

Numerous factors contributed to the rise of popular tourism in the United States, not the least of which were the discomfort and epidemic diseases that riddled the growing cities in summer. In 1826 the Erie Canal provided easy passage to Niagara Falls, which was reputed to have numerous inherent qualities and to radiate religious, hygienic, and intellectual blessings. Chauvinistic arguments for Grand-Touring America instead of Europe bloomed with the railroads.

As native poets and painters discovered the virtues of rural New England, their patrons and the socially elite came humbly after. The mob followed, further distorting the elevating intent of the original tour. Contemporary magazines are full of popular imitations of Coleridge, Wordsworth, Byron, and Bryant directed at New England landscapes.

Once in the field, a high proportion of the ordinary tourist's time was devoted to securing accommodations, conveniences, and safety. The only trouble with Yosemite Valley, according to a visitor in 1864, was the difficulty of finding out who won the election in Cincinnati. The most acute outer reality was composed of rates, schedules, and meals. When the passengers began wildly leaping from the train as it arrived at Niagara Falls in 1856, an observer "looked upon them at once as a select party of poets, overwhelmed by the enthusiastic desire to see the falls." As it turned out, "they were intent upon the first choice of rooms at the 'Cataract House.'"

Undoubtedly, the busywork of attending to trifles helped to avoid the unabridged emptiness between tourists and nature. Bryant's cathedral of the forest was theoretically as edifying as St. Paul's, and the crowd attended the outdoor temple as perfunctorily as it went to church on Sunday. The landscapes of "wild" nature in the heritage of ideas coming to us from Rousseau and the English artists and aestheticians were essentially rural. Most of Europe has been subjected to more than two thousand years of humanization through grazing, burning, cultivation, forestry, and "varmint" control. Even in the eighteenth century little evidence remained of European wilderness. Tourism was compounded of several factors, one of which was the idea of natural beneficence and the nobility of "wild" things. American tourists were unfortunate enough to have real wilderness close at hand. Neither the painters of the Hudson River school nor poets such as Bryan and Thoreau strayed far from the farms and villages of the principal river valleys. For them the Catskills and White Mountains were wilderness enough—and too much for the rank and file. It is not generally recognized that the flood of national sentiment for the preservation of Yosemite and Yellowstone had as much to do with their particular similarities to old humanized landscapes as any other factor.

Weaknesses in aesthetic and scientific discrimination made the tourist a ready prey to scenery vendors and fakers. Toll trails and a kind of promotional facade appeared in the landscape. As William Dean Howells put it with reference to Niagara Falls, "Their prodigious character was eked out by every factitious device to which the penalty of twenty-five cents could be attached." The Kaaterskill Falls, epitomized on canvas by Thomas Cole and Asher Durand and in the novels of J. F. Cooper and Washington Irving, were typically visited by fast coach in 1850. The carriage drew up before the Falls Hotel and the bar was opened immediately. As the tourists straggled out to the falls a boy was sent upstream to open a floodgate. It cost a quarter to have the falls turned on. When the crest had passed, the travelers returned to the bar and the coach, and thence to the Catskill Mountain House where dinner was always formal.

Travel was fraught with incredible haste long before automobiles entered the picture. Horace Greeley rode the 60 miles from Mariposa to Yosemite on one August day and had to be lifted down from his horse. As early as 1872 a visitor complained that "the Valley, in the height of its short season, is a confused scene of hurry, pushing and scrambling."

Paradoxically, there were long vexatious hours in the tourist's day that were not earmarked on the five- or eight-day tour program. Partly as a result of this, there was much measurement of the "curiosities" in height and volume, with favorable comparison to the better-known natural wonders of Europe. Calibrated walking sticks and plumb lines were part of a gentleman's equipment, as the "Claude glass" had been a century earlier. Tourists patronized "sybaritic" baths at the Cosmopolitan House in Yosemite, played cards, danced cotillions or bowled on the logs and stumps of Big Trees, and talked about whether it was more chivalrous or dangerous to chase butterflies on horseback. A howitzer was kept for making echoes. Since motors have overwhelmed the sounds of the landscape, making these pleasures vestigial, the howitzer has been replaced by an evening ritual in which a spectacular stream of burning faggots is pushed from a mountaintop.

Everyone is aware that improved transportation has not reduced the haste of the vacation. By the opposite route of the search for easy thrills, we approach the attitude toward nature advocated by St. Anselm in the twelfth century. Sin, he asserted, is proportional to the number of senses involved. As Edgar Anderson recently pointed out, few of us hear a landscape any more. Taste and smell are likewise handicapped. The day is here when the air-conditioned automobile carries us across Death Valley without discomfort, without disturbance to our heat receptors, and without any experience worth mentioning. We do not touch the burning sand.

The pursuit of comfort and amusement may be weighed against the degree of the encounter with nature. There is a choice to be made between capricious pastime and recreation. In a thoughtful essay, a Canadian geographer, Roy Wolf, suggests that the whole matter revolves about the question, "What relationship do I choose to assume with the environment?" The answer must be made from qualitative opposites: recreation and amusement; one an emotional debit, the other a credit. Whether he spends his spare time on a roller coaster or hiking in a forest may greatly influence the psychological equilibrium of the individual. To Wolf's conclusion it might also be added that the choice determines what happens to the environment. Incredibly, we campaign against litterbugs by trying to convince them that they should not exist—an approach entirely without success in the annals of pest control.

The history of tourism shows that we cannot account for the boobism of the tourist entirely as a machination of the twentieth century. All this century has done is narrow the lag between a vague desire for greater comfort and its realization. Perhaps the full swing of the pendulum will evoke a loathing out of satiation from which tomorrow's tourist will reject amusement. If and when such a millennium arrives, it may be too late. The improvers will have saturated the scene, leaving no landscape worthy of leisure.

The same park rangers who speak privately of tourists as "churnheads" return to their interpretive work annually, poorly paid and demanding though it is. It seems unlikely that many of them could do so with a conviction that they serve only as nuts and bolts in a booby machine racing nowhere. Is it fair to ask what are the redeeming qualities of tourism?

It should be noted first that the bulk of the summer travel in the United States is prodigious. It is unfortunate that almost nobody except builders of filling stations, chambers of commerce, and a few fitful figure-minded agencies have given the matter much attention.

The basic goodness of spare time is taken for granted by recreational agencies and labor unions. Ever since industrialization split the work week from the seven-day week there have been occasional objections to this assumption, of which Mr. Wolf's is a cogent example. The concept of leisure itself is fundamental, and from it any meaningful philosophy of tourism must extend. Leisure is reputed to have given us science and the arts. But a very few individuals are responsible for advances in these fields. Can we assume that the most meaningful aspect of leisure for all the other people would be merely appreciation or advance in amateur science and art? Perhaps even more significant to the plebeian is the effect of leisure on the whole range of his values, a spectrum that may be assigned to his religion. Josef Pieper has given us a modern restatement of leisure as "the basis of culture"—not, that is, as recuperation for more productive work, but as the antithesis to work. It is the only reason that labor is tolerable or meaningful. Yet leisure is not idleness. To Pieper it is not synonymous with spare time, as it is to the labor union, but a matrix for the evolution of the cultus, the basic religious body. The cultus is related to God through sacrifice, and from its rituals come the behavior and values of a society. Pieper compounds the Greek concept of leisure with Christian orientation in a fruitful way. It can be concluded that the use of leisure exhibits the essence of a particular society's relationship to the universe. The greatest contributions to civilization are made principally in leisure, not by "intellectual work" but by contemplative, intuitive, religious, or romantic orientation stemming from a release from travail.

Applied to the relationship between man and the landscape, it is apparent that the process envisaged by Pieper could lead to a world perception unfettered by a rigid thesis of nature for man's use. The ordinary citizen lives in the corridors of a narrow experience and knowledge. Its walls are his routine. In such circumstances his recognition of broad patterns in nature are not a rational product but a gift, unhinged from the necessary scheming in work or amusement. It is very fortunate that most tourists do not go afield, notebook in hand, but go instead with an undefined desire to see stimulating curiosities. This is the first step toward a recognition of themselves as part of nature.

We smile at the giddy antics of the self-styled, nineteenth-century "pilgrim" to the Falls or to the West, who suffers inconveniences in wild nature. But today the sacrifice is merely expressed in a different form—the splurger who pays outrageously for trinkets, meals, and gimcracks. The Christian ideas of suffering and sacrifice remain central to the idea of Western tourism. A unique feeling associated with certain encounters in nature was labeled "sublime" on the Grand Tour. It was a word transferred from religious to terrestrial description in the seventeenth century. Subsequently, God returned to the landscape. Persistently, though partly in fun and in part allegorically, a kind of equivalence appears in tourism between sensual reception and prayer, sacrifice and inconvenience, and worship and travel. There is an almost crusading zeal in the tourist. Of course, there were those who merely used the associated clichés and followed the stereotyped guidebook patter, but there was also John Muir who actually believed that God was visible in the landscape. The sublime was the awful and cosmic moment when one in a landscape perceived a relationship to God.

The pervasive and fascinating study of geology was a related influence in the evolution of tourism. The "pilgrims" flocking into Yosemite Valley, to Niagara Falls, or to the Yellowstone plateau gathered about peculiar geological phenomena; and so it has been since Richard Lassells interrupted his tour through Italian cities to peep over the lip of Etna as Petrarch had climbed Mont Ventoux. Almost everyone in the first half of the nineteenth century deliberated the stinging questions that geology raised about the Scriptures. Discussions concerning the Old Testament date of 4004 B.C.—for the creation of the world, the location of the Garden of Eden, the Great Flood, fossils, holocausts, cataclysms, and the permanence of hills and ocean basins—buzzed like angry bees. The naturalist was an amateur and everyone was something of a naturalist. Each considered the inconsistencies between bible and landscape from personal observations and reading.

Perhaps today's amateur naturalists are more apt to look at a book about mountains than at the mountains themselves. If they observe the mountain and read the book, the effect may be much stronger than one would expect from a kind of laboratory demonstration. For one thing, they reenact the discovery of the mountains as they were perceived by geologists, a procedure no different from rehearsing the discovery of the solar system as it was perceived by astronomers. The sublimity of these experiences reminds us that part of Galileo's "technique" was prayer. Perhaps because of our enthusiasm for history, or perhaps for more atavistic reasons, we never tire of putting ourselves on the stage to recreate great events. The traveler, too, recapitulates a cultural experience, giving his leisure a purpose and a liveliness and excitement otherwise absent.

William Gilpin, an eighteenth-century English vicar, wrote a book telling how to enjoy the English landscape. Since there was no completely wild nature remaining (that is, sublime wilderness), it was necessary to do two things while hiking through a woodland. The first was to imagine yourself to be lost, which might possibly release an instinctive fear.

The second necessity, according to the vicar, was to imagine that you were the first ever to set foot in that place. This precept may embody the most important aesthetic aspect of tourism: a reenactment of exploration. The relationship of hunter and fisherman is also a recreation of an earlier challenge. A myth of need and hunger is supported in the act of eating the catch. Gathering, husbandry, and hunting are recapitulations of a different class from the portrayal of discovery—which was itself a product of leisure. Whether this makes any difference in the perceptive value of its execution is doubtful.

The late Aldo Leopold believed that such mimo-drama had "split-rail" value, a virtue diminished by the use of gadgets. Split-rail values do not necessarily lead to perception, as when little boys play Cowboys-and-Indians. But they may catalyze meaning for the traveler in a landscape. When the airlines were young, passengers were given guidebooks for reading the landscape below. In those days, every passenger was a Wright Brother. Flying for the traveler has since become too high, fast, and monotonous; it is no longer an adventure. Perception may require an agent, whether a book or a personal interpreter. The federal government has assumed this obligation in the national parks with its publications, museums, and ranger-naturalists. There are various levels of the interpretation of nature, the most elementary being the identification of forms. Thousands of birdwatchers over the United States keep "life lists" and concentrate on recognition. Genuine perception proceeds beyond this: first into the life history of the identified organism or geological form and then into its relationships within a larger field. The last is much more difficult to perform, but it yields the only significant comprehension of patterns.

There are reasons why the interpretation of the landscape should be easier than it was a century ago. One is that more travelers live in cities. This may seem to be a paradox, but the city man, who is normally further removed from nature, is willing to accept a skillfully presented scheme of nature woven on a framework of pattern and process, even though he may not be convinced that it is important. The farmer will only accept parts. Another reason is that we know more about these patterns than we did a few years ago. A course in ecology was given recently to college seniors, none of whom were biologists, using the landscape as a focal point. Orientation and textbook emphasized process, pattern, and order versus disorder.

It is noteworthy that this course could be given in a liberal arts college, which represents the best surviving arena of the leisurely ideal. Visual habits in the normal workaday world of the public almost preclude the possibility of seeing whole patterns in nature. Christopher Fry pointed out several years ago that we learn very early in our lives the meaning of certain "signs" in our environment. While these may be important in our orientation, they are not looked at very closely after childhood. Most of us retain a small portion of childlike vision toward the part of the world that is economically unimportant, though we seldom bother to exercise it. For this reason we are much more likely to look closely at a landscape painting than at a landscape and we become better critics of the canvas than of the view. The acme of this form of blindness is expressed in Santayana's explanation of romanticism: "The promiscuous natural landscape . . . has no real unity, and therefore requires to have some form or other supplied by the fancy." Recent work in animal behavior has validated Fry's remarks with respect to "signs."

The best modern alternative to college life as a greenhouse for perception is the vacation, degraded, perverted, and streamlined though it may have become. The sportsman, observing the pattern of stream life in which moves the fish, is himself moving in a sphere that has no immediate connection to the conduct of business. The beauty of this activity is that it is completely useless. It has the merit of a low amusement but high reenactment quotient. Metamorphosis comes with the desire for an interpretation of the life of the fish. The whole watershed then assumes a new significance. Sometimes the sportsman appears before legislative committees urging antipollution laws, although neither the passage of laws nor any other particular objective need be the end of perception.

Tourists follow a classical pattern. They move through novel landscapes with minimal sign value. Detachment from daily niches increases the potential for understanding relationships. If amusements are avoided, one may even become curious. Going west is like replaying the history of the pioneer and the discoverer. Even in a city, if it is strange, wilderness with the charming possibility of becoming lost presents itself. The journey is at once a pilgrimage and a retreat. In this plastic and receptive mood the tourist is essentially a new and different person. Travel is broadening not because of the nature of travel but because of the traveler.

In 1873, Grace Greenwood prophesied a time when Yosemite Valley would contain "horse railroads and trotting tracks, hacks and hackstands, Saratoga trunks and croquet parties, elevators running up the face of El Capitan, the Domes plastered over with circus bills, and advertisements of 'Plantation Bitters.'"

Some observers may feel that this, essentially, has already come to pass. Others may wonder why it has not. When the tourist with her precious but plucked flowers is rudely rebuked by the ranger, it could mean that she reveals the dual nature of the traveler in the most provoking way.

Whatever Happened to Human Ecology?

T HE TITLES ON THE "NEW BOOK" shelf in the library suggest
that we stand on the threshold of a new holistic vision. One sees
there, for example, *The Ecological Perspective on Human Affairs* (Sprout
and Sprout, 1965). Upon examination, however, its only reference to
nature is found to be an attack on environmental determinism. *The
Ecology of Public Administration* (Riggs, 1962) would seem to be a Bible
for some future secretary of the interior, but it proves to be only a
political text. We are attracted by a mysterious title, *The Discipline of
the Cave* (Findlay, 1966). Is it Neanderthal ecology? It is an epistemology having nothing to do with caves. Still another, subtitled *Pastoral Ecology,* is about the ministry instead of men and ungulates. It is
evident that ecology has become an "in" word. Environmentalism

swings. What is the half-life of fashionable words and what is their connection, if any, to the general intelligence?

There is increasing general ecological awareness. Air and water pollution are often considered in such a context. But there is also a strong cultural resistance to the idea as a whole (White, 1967). Even Thomas Huxley's famous book *Man's Place in Nature* (Huxley, 1863), dealt with the question only of descent and morphology and set a precedent by which two generations of students of "man the animal" could elucidate primate anatomy and ignore primate ecology. The reaction of the conventional wisdom against the theory of evolution in the nineteenth century was against its phylogenetic implications. Now it seems that much of the anguished a priori opposition to the biological creation and natural lineage of humans has been redirected against ecological interdependence that links the living biota with humankind.

Development of Ecology

Efforts to include humans in ecology took several directions as ecology itself developed in the United States. By 1930, for instance, Victor E. Shelford was putting two decades of reflection to work at Illinois with graduate seminars on the subject of human relationships to the environment. The framework was taxonomic community, dominance, succession, symbiosis, and niche. The subject was characterized by W. C. Van Deventer as "the man-dominated/man-influenced community of civilization" (Shelford, 1935). The idea is carried forth today as landscape ecology by Shelford's students. Van Deventer's course at Western Michigan University, for example, "Human Ecology: A Study of Man in Relation to His Living Environment," is a straightforward account of our familiar world as the product of human activity and ecological processes. It is a familiar theme of other ecologies as well, such as Edgar Anderson (1952) and Frank Fraser Darling (1955).

Meanwhile, the urban environment was scrutinized by sociologists. Robert Ezra Park (1936) explains that it was an attempt to operate between geography and economics and to apply to the interrelations of human beings the type of analysis developed in the study of plants and animals. He and E. W. Burgess are said by Hawley (1950) to have coined the phrase "human ecology" in 1921, though it was also pioneered in the United States by R. D. McKenzie.

Though biologists may be gratified by the vigor with which the sociologists embraced ecology, they are likely to read their papers with some uneasiness. Just as the demographers got the uses of "fertility" and "fecundity" backward, the sociologists took a terminology devised originally for describing interspecific processes and applied it wholly within the human "community." From "dominance" and "symbiosis" to "territory" and "competition," nothing meant quite what it had. Forty years later, "human ecology" has become that branch of sociology dealing largely with urban geography. It has spawned its own raft of texts that testify to its essentially analogical character and restricted emphasis, narrower and less biological than Park's own original views. Some examples of the nature and scope of studies in this tradition may be seen in a current anthology (Theodorson, 1961).

Ecological Spin-Off

The tendency for ecology to decentralize—to spin off independent systems and ideas—linked it to cultural demography by way of animal population studies, which continue to be one of the most vital areas in ecology. If these studies do not startle the actuarial world they at least reaffirm the reality of population limitation. What *Fortune* magazine was still calling "the Malthusian Mischief" (Winocour, 1952) was stirred up by a few outspoken ecologists and like-minded humanists in the years after World War II. (The most cogent statements on human population ecology are probably those of Aldous Huxley published in 1950 and 1963).

Although human population saturation was not a new idea, it was still virtually subversive to declare that humans must limit their numbers. *The Road to Survival* (Vogt, 1948) is a case in point, the contribution of a biologist to the creation of an awareness of population density. The biologists were strident when stridence was needed, while urbanity has lately marked the dialogue in the manner, for instance, of Edward Deevey's (1956, 1960) factual but far from feverish essays. Animal behavior is another active field with roots in ecology. Its implications for the human situation are great. And great is the peril for any who attempt the bridge, for it is guarded not only by bulwarks of nonscientific indignation, but by scientific critics prepared to cut down anybody who is not as cautious as themselves in translating the data from animal social life. Witness their treatment in reviews of *The Territorial Imperative* (Ardrey, 1966) and *On Aggression* (Lorenz, 1966).

There is a large body of cautionary literature stimulated by ecology that can only be suggested. It ranges from the preservation of species and species diversity (Elton, 1958) to the status of forests, water, and soil (Lowdermilk, 1952) and to the preservation of processes pertinent to the whole biosphere (Cole, 1966). The cautionary literature has been criticized (as a standard rejoinder by vested interests) for its pessimism, determinism, and primitivism. In the face of the destructive opportunism and technological fluidity that threaten to change the environment substantially before its ecology can be understood, most ecologists are discouraged. The forces poisoning the air and felling the trees around them are not merely the expedients of local self-seekers, but the visible front of a Cartesian-Hobbesian worldview. In such a society the reaction to *Biology or Oblivion* (Hocking, 1965) is the frown of disapproval, for it adds the insult of ultimatum to the objectionable cry of Cassandra. A possible solution is implicit in the advice of Alexander F. Skutch (1946) that we must go forward to Nature, not back to it. The ecologists' ability to untangle the meaning of that statement will determine their influence in the planning explosion that lies ahead.

Diverse other fields also make their humanizing claims on ecology, with occasional attempts at reconciliation (Quinn, 1940). "Geography as Human Ecology" (Barrows, 1923) is intermediate in time and by token representative of a range of interests going back to the nineteenth-century environmentalists and coming forward to the current outbreak of ecology in regional and urban design. Interest by historians in ecology is epitomized by the work of James Malin (1950). Anthropologists such as Laura Thompson (1949) and Betty Meggers (1954) represent a group of social scientists who do not believe with L. A. White (1949) that culture can be explained only in terms of itself. A recent symposium on man and animals (Leeds and Vayda, 1965) gives the impression that anthropology may be the generator of a true scientific discipline of human ecology in the future. Medical studies ranging from the ecology of the skin (Marples, 1965) to dissatisfaction with medical convention about cause and effect in disease (Dubos, 1966) are current topics indicating the inseparability of certain kinds of physiology from ecology, tracing from the systems theory of Claude Bernard (1878–1879) and pursuits of epidemiology (May, 1958). Other examples of this active field of research are to be found in current anthologies (Bresler, 1966) and in handbooks of physiology (Dill, 1964). The physiological approach to population densities (Christian and Davis, 1964) contacts the study of human stress (Selye, 1956) on one side and territoriality (Wynne-Edwards, 1962) on the other.

Ecology and Man

If ecology applied to man seems to radiate into all aspects of human existence, its general principles become correspondingly more important. Its central concept is the ecosystem: the pathways of element, energy, and information flow. In controversial matters such as chemical pest control, the ecosystem idea is a useful framework for sorting the tangle of claims and counterclaims (Egler, 1964). The framework may emphasize the biotic pyramid (Leopold, 1948), biogeochemical cycles (Hutchinson, 1949), or energy in the ecosystem (Sears, 1959).

It seems that there are ecologies of man, but no definitive ecology in the sense that we have an ecology of the robin or the pine tree. The situation is not altered by the large body of general perspectives on people in the biosphere. Such comprehensive views, like those of C. C. Adams (1935), Lee Dice (1955), Marston Bates (1953), May Watts (1957), Sir George Stapledon (1964), and others attest to a degree of predictability of the effects of human actions on other species and the environment. The effects of the destruction of a pine forest, for example, on the local microclimate, water supply, and biota can be seen and are crucial in the future of mankind. But the limitation of general ecology is that it throws no light on the question of whether humans actually will or will not cut down the pine forest. Any number of other disciplines—humane, social, and scientific— would seem to have some part of an answer to that question, and the natural scientist's immediate reaction would be to leave the question to them.

The broad aspects of human ecology cannot be subdivided according to academic tradition. Committees seldom bridge interdisciplinary problems, but only confront each other at the borders. It is apparent that individual biologists and others must explore the human/nature interpenetration as a whole: as a cybernetic or functional system (Collins, 1965).

Abandonment of an Idea

It may be anticipated that one characteristic of such syntheses will be abandonment of the idea that culture is a superstructure. Led by Sir Julian Huxley (1948), biologists themselves were largely responsible for a view of human life as two-layered—one composed of vegetative functions and genetic evolution, the other, superimposed upon it, shaped independently by tradition and learning. Nonbiologists fastened only too gladly on a concept that freed them from having to learn biology and their special fields from what Albert Schweitzer called animal tracks on the clean kitchen floor of theology (Joy, 1951).

The general cultural resistance to the ideas of ecology is, in effect, a repudiation of the natural world itself. The answers to the question of what will be done with a pine forest are to be found in the working of history and prehistory, emerging from systems of psychic, biological, and environmental factors. "Cut the forest, there is no meat in it!" whispers the australopithecene in our brain; "Cut the forest, there is no grass in it!" murmurs the pastoral memory; "Cut the forest, its soil needs sunlight!" cries the shade of the primeval farmer; "Cut the forest, for the wilderness is pagan!" thunders the Church; "Cut the forest, it blocks my view of the city!" cries the humanist; "Cut the forest, for the wood we can get!" pleads the haggler; "Cut the forest, to prove that ours is the choice!" howls the existentialist. All are of one mind and many minds. In all, the passion is complex—a swirling current of unconscious symbol and rational decision, of the most delicate feeling inseparable from the imprint of growth and perception. To cut the forest or not to cut it reflects a personal and social cosmology, not any aspect of which is free from the experience of gravity, earth, sky, air, sunlight, or the peculiar individual quirks of the unfolding of life in a natural world.

The Central Problem

Perhaps the central problem of human ecology may be characterized as the relationship of the mind to nature. Such a view was lucidly advanced by Sterling Lamprect (1938). His essay "Man's Place in Nature," the same as Huxley's title, added thought to the phylogenetic and the ecological web-of-life perspectives. Some examples of the insight generated by such minding are to be found in the work of a biologist, an art historian, a theologian, a mythologist, and some anthropologists.

In *The Enchanted Voyage* (Hutchinson, 1962), a web of sea imagery is explored as an extension from terrestrial mythology. In the traditions of European seafaring, the voyage is the discovery of forms and processes that seem to echo and yet to enlarge those of the land. Its symbols of fertility, the life cycle, and death not only enriched art and imagination of Europe but stimulated the motive of discovery so that even the contemporary oceanographic voyage is a search in and for enchanted worlds. The study of such matters, in this instance by a biologist, illuminates the interpenetration of man and nature in such a way that we see mythologizing as a form of the perception of nature.

The Earth, the Temple, and the Gods (Scully, 1962) examines the relationship of landform and religious architecture—which is inseparable from the ritual ceremony of early Greek temples. The Greeks have been generally looked upon as insensitive to the beauty of the landscape painting or the literary celebration of wild solitude. But Scully goes beyond these derivatives of the modern idea of scenery to show that the Greek response was not so abstract or superficial. The temple was built literally in the lap of the goddess, whose immediate local features were embodied in the morphology of the natural locale. The sanctuaries thus completed by architecture create a landscape art in which mind and physiography are inseparable—part of the feedback of a religious ceremony. Like the animal art of Paleolithic cave dwellers, the Greek temple sites are fossils of the functional genesis of art—so different from its modern civilized descendants, kept in our time by patrons as relict, domesticated curiosities for display or vicarious excitement.

The sacredness of space has been the subject of study elsewhere also. It comes at a time when the world is becoming notably homogenized. "The nonhomogeneity of space," says Mircea Eliade (1959), "is a primordial experience, homologizable to a founding of the world. It is not a matter of theoretical speculation, but of a primary religious experience that precedes all reflection on the world." The process of homologizing is explored in a brilliant book by Joseph Campbell, *The Masks of God: Primitive Mythology* (1959). His thesis is derived directly from ethological concepts of the sign stimulus and of imprinting. He identifies the "imprints of experience"—such events as birth, waking, sucking, perception of the face and sun—as structuring forms upon which ceremony and ritual build a wider reality. The ceremonial use of mask and environment of the Paleolithic cave rituals, for example, return the participating neophyte back into the womb to be reborn. The mask and pictures are supernormal recreations. He is initiated into a metaphysical schema that is homologized by a reenactment of his own biological experience.

In an essay of great insight and beauty, Edith Cobb (1959) reports on her identification of "a special period, the little-understood, prepubertal, halcyon, middle age of childhood, approximately from five or six to eleven or twelve—between the strivings of animal infancy and the storms of adolescence—when the natural world is experienced in some highly evocative way, producing in the child a sense of some profound continuity with natural processes and presenting overt evidence of a biological basis of intuition." The development of the perceptual processes is inseparable from the child's activity in nature—an essential background to the ultimate and unique human types of creativity. She sees the "embryology of mind" as a genuinely ecological process and views genius as recall based on the "child's early perceptual continuity with nature." This "deep need to make a world the way the world was made" is "the only truly effective counteragent to the forces of internal conflict which until recently were considered the major subjects of study, the main background to purpose in life." And she calls for a "redefinition of human relations in terms of man's total relations with 'outerness,' with nature itself."

The New Ecology

The new kind of ecology represented by Campbell and Cobb indicates that problems of identity and delinquency are more than social symptoms that will be solved by better urban planning or even by "worthwhile" vocations (Goodman, 1956). The processes of which they speak must be affirmed and nourished. The "crises of adolescence" can be like acute vitamin deficiencies. Among the Pueblo Indians, by contrast, "the tribe knew that its hold upon the future, the persistence of its tradition, of its religion, of its emotional orientation, of its ancient soul which involved the world-soul, were dependent on the adolescent disciplines" (Collier, 1931). Building upon and coming after the "middle age of childhood," the early adolescent years fix the individual destiny. The Pueblo boys, returning to the community after a year's secret training, are "radiant of face, full-rounded and powerful of body, modest, detached: they were men now, keepers of the secrets, houses of the Spirit, reincarnations of the countless generations of their race; with 'reconditional reflexes,' with emotions organized toward their community, with a connection formed until death between their individual beings and that mythopoeic universe—that cosmic illusion—that real world—as the case may be, which both makes man through its dreams and is made by man's dreams."

Should our society settle for less? "The difference between primitive and civilized mentality is not absolute; there is no chasm between them, as some scholars have thought. We live in the same world as the savages. Our deepest experience, needs, and aspirations are the same, as surely as the crucial biological and psychic transitions occur in the life of every human being and force culture to take account of them in aesthetic forms" (Chase, 1949).

A final example is a group of studies that correlate anthropological, paleontological, and primate ethological evidence on the evolution of hominids leading to present-day humans (Washburn, 1961). On the whole, it seems to mark a return to a more fruitful exploration of instinct. It supports Julian Steward's (1936) contention that a large part of the human emergence, attitude toward nature, personality, and social organization was shaped by the conditions of social carnivorousness. The idea that the Pleistocene might hold the clues to the nature of man is not new, but the meaning of that experience in terms of the kinds of environments which men create, go to, and dream of remains to be discovered. Probably the reaction against the idea of human instinct (qua determinism) over the past seventy-five years crippled our awareness of the richness of innate human behavior and its interplay with learning. Now biology and psychology must confront the ecological function of the unconscious and anthropoid and primitive life as a frame of reference—in short, acknowledge the place of the primate soul and the hunting heart at the center of "the phenomenon of man."

Destiny

One might conclude that the destiny of human ecology is to accept its own eclectic nature. It would be impertinent to attempt to define it now so as to exclude its historical forms or its descendant and peripheral disciplines. The examples I have given, illustrating the kind of thinking and synthesis that seem to fulfill the great promise of human ecology, are not research reports. They are works of art. Human ecology will probably never become the nucleus of a graduate department or government agency, except perhaps in one of its more restricted senses. There are at least three general approaches: a kind of extended individual physiology of the sort stimulated by space flight research and stress syndromes; the implications for man from general landscape and ecosystem ecology; and, finally, the exploration of nature and the human mind as a feedback system. Human ecology has no sacred core to guard from Philistines. It will be healthiest perhaps when running out in all directions. Its practical significance may be the preservation of the earth and all its inhabitants.

REFERENCES

Adams, C. C. "The Relation of General Ecology to Human Ecology." *Ecology* 16(3) (1935):316–335.

Anderson, Edgar. *Plants, Man and Life.* Boston: Little, Brown, 1952.

Ardrey, Robert. *The Territorial Imperative.* New York: Atheneum, 1966.

Barrows, Harlan. "Geography as Human Ecology," *Annals of the Association of American Geographers* 13 (1923):1.

Bates, Marston. "Human Ecology." *Anthropology Today,* ed. A. L. Kroeber. Chicago: University of Chicago Press, 1953.

Bernard, Claud. *Leçons sur les Phénomènes de la Vie Communs aux Animaux et aux Végétaux* . Paris: J. B. Bailliere et Fils, 1878–1879.

Bresler, Jack. *Human Ecology: Collected Readings.* Reading, Mass.: Addison-Wesley, 1966.

Campbell, Joseph. *The Masks of God: Primitive Mythology.* New York: Viking, 1959.

Chase, Richard. *Quest for Myth.* Baton Rouge: Louisiana State University Press, 1949.

Christian, John J., and David E. Davis. "Endocrines, Behavior, and Population," *Science* 146 (1964):3651.

Cobb, Edith. "The Ecology of Imagination in Childhood." *Daedalus* (Summer 1959):537–548.

Cole, Lamont. "Man's Ecosystem." *BioScience* 16(4) (1966):243–248.

Collier, John. "Fullness of Life Through Leisure." In *Mind-Body Relationship,* ed. J. B. Nash. New York: Barnes, 1931.

Collins, Paul W. "Functional Analysis." *Man, Culture and Animals,* eds. Anthony Leeds and Andrew P. Vayda. Publication 78. Washington, D.C.: American Association for the Advancement of Science, 1965.

Darling, Frank Frazer. *West Highland Survey: An Essay in Human Ecology.* London: Oxford University Press, 1955.

Deevey, Edward. "The Human Crop." *Scientific American* 194(34) (1956):105–112.

————. "The Human Population." *Scientific American* 203(3) (1960):195–204.

Dice, Lee Ramond. *Man's Nature and Nature's Man: The Ecology of Human Communities*. Ann Arbor: University of Michigan Press, 1955.

Dill, D. B. *Handbook of Physiology. Section 4: Adaptation to Environment*. Washington, D. C.: American Physiological Society, 1964.

Dubos, Rene. "Second Thoughts on the Germ Theory." *Scientific American* 192(5) (1955):31–35.

Egler, Frank. "Pesticides—in Our Ecosystem." *American Scientist* 52 (1964): 110–136.

Eliade, Mircea. *The Sacred and the Profane*. New York: Harcourt Brace, 1959.

Elton, Charles. *The Ecology of Invasions by Animals and Plants*. London: Methuen, 1958.

Findlay, John N. *The Discipline of the Cave*. London: Allen & Unwin, 1966.

Goodman, Paul. *Growing Up Absurd*. New York: Random House, 1956.

Hawley, Amos. *Human Ecology: A Theory of Community Structure*. New York: Ronald Press, 1950.

Hocking, Brian. *Biology or Oblivion*. Cambridge, Mass.: Schenkman, 1965.

Hutchinson, G. E. "On Living in the Biosphere." *Scientific Monthly* (December 1948):393–395.

————. *The Enchanted Voyage*. New Haven: Yale University Press, 1962.

Huxley, Aldous. *The Double Crisis: Themes and Variations*. New York: Harper, 1950.

————. *The Politics of Ecology*. Santa Barbara, Calif.: Center for the Study of Democratic Institutions, 1963.

Huxley, Julian. *Man in the Modern World*. New York: Harper & Row, 1948.

Huxley, Thomas Henry. *Man's Place in Nature*. London, 1863.

Joy, Charles R. *The Animal World of Albert Schweitzer: Jungle Insights into Reverence for Life*. Boston, Mass.: Beacon, 1951.

Lamprect, Sterling. "Man's Place in Nature." *American Scholar* 7(1) (1938): 60–77.

Leeds, Anthony, and Andrew P. Vayda, eds. *Man, Culture, and Animals*. Publication 78, Washington, D. C.: American Association for the Advancement of Science, 1965.

Leopold, Aldo. *A Sand County Almanac*. New York: Oxford University Press, 1948.

Lorenz, Konrad. *On Aggression*. New York: Harcourt Brace, 1966.

Lowdermilk, W. C. *Conquest of the Land Over 7000 Years*. Bulletin 99. Washington, D.C.: USDA, 1952.

Malin, James C. *The Grasslands of North America*. Ann Arbor: Edwards Brothers, 1947.

Marples, Mary. *The Ecology of the Human Skin*. Springfield, Ill.: Thomas, 1965.

May, Jacques M. *The Ecology of Human Disease*. New York: MD Publications, 1958.

Meggers, Betty J. "Environmental Limitation on the Development of Culture." *American Anthropologist*. 56 (1954):803–804.

Park, Robert Ezra. "Human Ecology." *American Journal of Sociology* 42(1) (1936):1–15.

Quinn, J. A. "Human Ecology and Interactional Ecology." *American Sociology Review* 5 (1940):713–22.

Riggs, Fred Warren. *The Ecology of Public Administration*. New Delhi: Indian Institute for Public Affairs, 1962.

Scully, Vincent. *The Earth, the Temple, and the Gods*. New Haven: Yale University Press, 1962.

Sears, Paul B. "The Steady State: Physical Law and Moral Choice." *Key Reporter* 34(2) (1959).

Selye, Hans. *The Stress of Life*. New York: MacGraw-Hill, 1956.

Shelford, Victor E. "The Physical Environment." *A Handbook for Social Psychology*, ed. Guy Murchison. Worcester, Mass.: Clark University Press, 1935.

Skutch, Alexander F. "Back—or Forward—to Nature?" *Nature* 39(9) (1946): 457–460.

Sprout, Harold, and Margaret Sprout. *The Ecological Perspective on Human Affairs*. Princeton, N.J.: Princeton University Press, 1965.

Stapledon, Sir George. *Human Ecology*. London: Faber, 1964.

Steward, Julian H. *The Economic and Social Basis of Primitive Bands: Essays in Honor of Alfred L. Kroeber*. Berkeley: University of California Press, 1936.

Theodorson, George A. *Studies in Human Ecology*. New York: Harper, 1961.

Thompson, Laura. "The Basic Conservation Problem." *Scientific Monthly* 68(2) (February 1949):129–131.

Vogt, William. *The Road to Survival*. New York: Sloane, 1948.

Washburn, Sherwood, ed. *Social Life of Early Man*. Chicago: Wenner Gren, 1961.

Watts, May. *Reading the Landscape*. New York: Macmillan, 1957.

White, L. A. *The Science of Culture*. New York: Farrar, Straus, 1949.

White, Lynn, Jr. "The Historical Roots of Our Ecological Crisis." *Science* 155 (1967):3767.

Winocour, Jack. "The Malthusian Mischief." *Fortune*, May 1952, pp. 96 and 212.

Wynne-Edwards, V. C. *Animal Dispersion in Relation to Social Behavior*. New York: Stechert-Hafner, 1962.

 # The Conflict
of Ideology
and Ecology

I N RECENT YEARS there has been a concentrated effort in many disciplines to locate those ideas and sentiments in history that underlay the exploitation and destruction of ecosystems. This widespread riffling of the documents and institutions upon which modern presuppositions about the relationship of humans and nature are founded is preceded by a less introspective explanation based on the "gospel of efficiency," a creed espoused by the early conservationists that postulated the origin of the difficulty in poor laws, bad management, inadequate planning, and wasteful habits. Predicated on the controlled expansion of an economy geared to ever more efficient conversion of ecosystems and habitats into jobs and products, it was a view that sat well with all the major institutions of government and business. Indeed, in America conservation was conservative: a rich man's hobby and big foundation playground.

When doubts arose about these premises and it began to appear that deep weaknesses in the system and not simply the system's priorities were at fault, there was an academic scramble to open up the history of ideas, doctrines of theology, origins of aesthetics, and the confines of "linear thought" to lay bare the root causes of our environmental dilemmas. Out of this rich lode came monographs on the theory and practice of the domination of nature, Greek and Roman hubris, the arrogance of patriarchal, pastoral civilizations under whose feet the world was washing away, Judaic historicism and Christian otherworldliness, the mechanizations of science, and illusions of homocentrism. It was a big rock that scholarship turned over and many of its cryptic denizens hunched there have yet to be collected. It is apparent that, like ecosystems, the web of our perception of nature ramifies without end, pervading thought and value in all the arts and sciences. In all fields it has become possible, even somewhat fashionable, to attend to those components heretofore ignored.

One of the consequences of this new exploration is that C. P. Snow's widely heralded "two-cultures" version of the modern world breaks down. Science does not always compel a vision of nature different than that of the humanities. Both are strongholds of the hierarchic view of the world, mankind as the privileged species, and the subordination of the nonhuman for the benefit and exaltation of people. Conversely, both contain minorities in which a more organic model and more humble philosophy prevail.

Another consequence is the ways in which science influences the nonsciences, and it is to one of these that I wish to draw attention. It occurs as an aspect of the style of modern thought which, in effect, is detached and ostensibly nonjudgmental, not so much noncommittal as an extension of the view that all ideas have their merits and should be heard. There is nothing objectionable about it on the surface, but it is the strategy of the calculating intellect which enables individuals to deliberate on the assumption that they are restraining their personal preferences from matters of public discussion and are open to divergent views.

This cherished posture, in its modern form, is interconnected with the discovery of culture. That discovery occurred in the twentieth century as the social sciences realized that the world's ethnic groups do not constitute evolutionary stages which culminate in Western Europe, but are in fact equally valid ways of life. It is not this premise itself which is at issue but a corollary flowing from it, that appraisals must be withheld about the practices of other people because such valuing is biased and culture-bound. Carried one step further, this view is that, even within a single society, no absolute values are possible. As academics we live in a world dominated by the fallout from this relativism which flourishes in a multinational political environment but castrates the educational process by depriving it of any commitment except to its own unflinching pliancy.

Oddly it is a style inextricably connected to the myth of ideology. Nothing might seem farther from the militant assertion of one's own beliefs than the determined flaccidity of commitment I have described. But the myth of ideology is that the human animal is engaged in a lifelong process of choosing sides, mobilizing conviction in dialectics, collating opinions, dipping in and out of beliefs, and, in short, creating a concept of the self in terms of selective preference. To that end the academics and intellectuals see themselves as the agents of internal and external dialogues, committed only to the ideal of identity by belief and will.

Although one is identified with a "position" in such a culture, its actual basis is incipient abandonment of positions. The individual articulates a thesis of identity the way a lawyer takes a case, as though it were detachable: One may take the opposite side tomorrow from a hotly defended position today. The myth holds all those elements constituting the self or group as equally detachable: politics, social role, "lifestyle," gender, and relationships in a wider ecological context. But the ease of detachment is merely a supposition, and it is quite possible that we are not that kind of animal at all. If so, attempts to change ecological relationships may be very different in consequences than detachment from a legal position. The kinds of damage that are possible in exercising this philosophy of casual disengagement have not been much studied.

Today every field of the study of mankind is deeply committed to environmental relativism. Even the concept of "environmentalism" has become a psychological doctrine denying inherent constraints on the human organism. Geography, with its blasé endorsement of economic determinism; history, our study of the "rise" of civilization on a Promethean theme; the arts, which have separated abstract qualities from content; anthropology and sociology, with their social process and ethnological cataloging on the theme of "everything's possible"; and the natural sciences, with their posture of value-free fact-finding—all seem to confirm that "man makes himself" no matter how the world is made.

I am fully aware of the crippling effect on thought of the unwillingness to consider alternatives or the indulgence of whim and emotion. To turn the mind over to the arbitrary acts of irrational behavior and rationalizing mindlessness would be to surrender qualities that are precious, rare, and hard-won. Tyranny lies in that direction. And yet we are engaged in a worldwide devastation of ecosystems and other species on the grounds that people are related to the natural world only by "paradigms," "models," "schemata," or other constructs that are at best variations on archetypes and at worst fickle opinions shared by a group. Thus the biological concept of evolution is seen not as a scientific formulation of the integrative processes of nature but as one more construct in a marketplace of ideas. Myths are viewed in a framework of comparative mythology, prose as comparative literature. Any and all aspects of human life assume this mantle of footless disposition: the human population, the existence of other species, the use of soils, forests, and waters, the manipulation of watersheds, the regulation of energy and nutrient systems, the fate of public lands and public air. Each of these becomes in this mode of thought a bundle of "issues" in which coalitions of political power rise and fall, working out their compromises on the shared assumption that the world has no inherent structure, no given, only an order projected upon it that makes it seem coherent. One chooses ecological relationships the way one chooses a political party or brands of groceries.

As unrealistic as this attitude may seem, it is the position of modern scholarship. It is the intellectual manifestation of the Faustian premise of human destiny by the assertion of will and the domination of nature—stoutly defended on the grounds that (1) humans are a special case and therefore free to make a world according to their desire and (2) that the notion of "determinism" (taken to mean any constraint by the nonhuman) is a cold hand on imagination, creativity, or the human spirit and is inevitably abused to justify injustice.

On the grounds that one culture is intrinsically no better than another, we are therefore expected to accept, for instance, that goat pastoralism is no worse than any other land use, and that it cannot, in fact, be judged by members of outside cultures. Consequently we have created a free-association fantasyland for every area of thought touching on either the "cradle" lands of civilization in the Middle East or on the present "Third World." Back-to-the-land movements, bearing ecological banners and celebrating simplified "lifestyles," are widely adopting the goat on the misapprehension that all the attributes of preindustrial agriculture are ideological.

Because it has given the world a five-thousand-year lesson in environmental catastrophe, one might expect that goat keeping and its effects would represent a generally understood and acceptable principle. But the goat remains immune. It is sanctioned on government lands, given as foreign aid, praised by Peace Corps and missionaries, idealized by literary enthusiasts for classical allusions, defended by geographers, and cuddled by subculture groups who refuse to accept that any animal so cute and sociable is not a friend to humans.

This is but one example in the whole realm of human/nature relationships. Lacking a true cultus to undergird our thought and discussion, we have instead a cult of relativity with its glaze of humane affirmation of brotherhood: the unwillingness to commit beyond dialects on the grounds that personal choice and cultural context are the only absolutes. Fearful of appearing provincial or prejudiced, the intellectual style holds only to the right of equivocation.

The observation that a society based on cultural relativism inexorably extends its withdrawal of judgment and appraisal to all of its areas of interest, including human relationship to nature, precipitates us into a serious double-bind. How can we divest ourselves of that autotelic posture which says in effect that there is no evolutionary context of human life, no inherent constraints on our freedom to do as we please in the natural world—to throw off that hubris in which the world is viewed as inert stuff for our manipulation and still retain our tolerance for the views of other peoples? If peace in the world is to be based on mutual respect, a willingness to live in a world of many races and many customs, how can we avoid the accommodative attitude I have just been criticizing? How can we be at once committed and yet affirm the diversity of peoples?

In short, the advocacy of an absolute sense of human ecology and a relativistic one of human cultures seems contradictory. But I think it need not be. The conservationists have held that the attribution of things as resources was primarily a cultural designation. That may be so, but the reality of ecosystems and biomes, the interdependent community of species, the roles of soil and sea as the matrix of life, are not cultural concepts. They are previous to cultures. It may indeed be chauvinistic for one people to ridicule an alien notion that, for instance, the mistletoe, a shrubby parasite, is spiritually potent. But the extermination of whales or pollution of the stratosphere is another matter.

What is the difference? How can we forbear on the grounds of universal brotherhood and local customs on the one hand and be adamant about natural systems on the other? Within the context of modern existential relativism, there is no resolution of this dilemma. To those who accept all things on the grounds of the privilege of personal or cultural conviction, for whom reality is created wholly anew with each perceiving agent, and alike for those committed to the homocentric destiny in which mankind transcends the usual order of nature, the contradiction can only be resolved by the assertion that all such actions carry the rights of tribal hegemony. Carried to its logical end in a democracy of individualism, everyone's private vision is as valid as every other. This is precisely where academic, intellectualist thought has brought us.

If ecological thought—that is, natural philosophy or ecosophy— is seen as part of that neutral matrix to be shaped according to the circumstances and impulses of each tradition, then there can be no prior understanding. It means, in effect, that there is no such thing as the human species except as a kind of transmitter or container for fluid thought. It follows that there are no other species either, since, like us, they cannot be studied as a species but only as specific and particular instances. The parable of the blind men and the elephant becomes appropriate in a strange way: Reality depends on your per- spective and the occasion; therefore there is no elephant, only a word by which different individuals, groups, cultures refer to a kind of experience.

Geographer David Lowenthal asserted that the American con- cern for wilderness had nothing to do with insight into the founda- tions of our existence but was only a matter of taste—and rather poor taste at that. "The bison," he wrote, "is not a great shaggy beast. It is only a congeries of feelings."

The poor bison! It does not exist except in the multifold ways that different cultures choose to see it. In the same spirit, the typical trick for impressing freshmen in introductory psychology is to sug- gest that the tree falling on an island where there are no people makes no sound.

Thus things cannot have an enduring relationship with their environment if their reality depends on the contingent mind. Art, in this view, has no communicative function except by idiosyncratic chance. This capriciousness, or "the cult of eccentric originality" as Sibyl Moholy-Nagy calls it, is shared by the artists themselves, as when Le Corbusier or Soleri build cities or buildings unrelated to the surroundings. History itself is often seen as a collage of such extrane- ous acts—a transcendent creativity answering only to seemingly illogical whim.

Because the central question concerning the relationship of human to nature is the nature of nature, that is, of the Other, its reciprocal aspect is the nature of the self. All concepts of the natural thus turn on the definition of *Homo sapiens*. If humans are indeed members of ecosystems and not their god, then paleoanthropologists have indeed fostered our understanding of the genus *Homo* as an evolving, biological species. The mainstream of anthropology, however, clings tenaciously to the dazzle of comparative ethnology, the striving for cultural analysis pioneered by Boas and given brilliant recent articulation by Clifford Geertz, who says that since human behavior is only manifest in a particular culture, the underlying species-specific behavior is not only unavailable but virtually nonexistent. Thus has anthropology, like the other social sciences, surrendered its opportunities to define what it means to be human except by intraspecies comparisons—that is, by human difference. Humans may indeed have food habits, they seem to say, but all we can study are knives, forks, and spoons.

In this way a field with the brightest possibilities for defining human ecology yielded to the dialectics of ideology and the ideology of dialectics. Ideology, the master myth of our time, is the cloak of reasoned partisanship in a world where there are allowed only internally logical options. Within the frame of competing world religions (what Gary Snyder has called "the cosmopolitan religions"), isms, ologies, parties, sects, and a dime-store counter of psychological trinket-theories of the self we seem to have settled for the deep premise that the only environment is a kind of ambiance or closet from which we choose today's reality costume.

The modern imagination, as Louis Halle has said, is ideological. Thought is so grossly confused with consciousness that a prominent psychologist can write on the "origin of consciousness" as an existential dilemma occurring in historical times, a view woefully deficient in what Robert Ardrey terms the sense of "four-dimensional man."

It has been widely observed that since about the time of Francis Bacon, the West abandoned its preoccupation with paradise past and fixed its attention on future utopias to be created by the reign of politician-scientists. Having turned our back on the past as the primal element in our identity, we entered a wonderland of possibilities. Liberated from fifteen hundred years of desultory regret and lament for the lost Eden, the energies of hope and optimism sprang up in the name of progress. It is not surprising that Darwin's theory of evolution has had so little impact on the modern consciousness. The Church may have fought it, but to the secular mind the image of eons of protohuman bestiality was no more attractive than a world degenerating since Adam. Only in one respect was Darwin preferable: He seemed to support the theory of progress, though of course evolution has nothing whatever to do with that concept.

This abuse of Darwinian thought may prove in the long run to be more destructive than social Darwinism. From the confusion of progress and evolution we derive the corollaries of the normalcy of perpetual change and the idea of adaptability. Shorn of their time perspectives these aspects of evolution appear to justify change for its own sake, unlimited human flexibility, and therefore the commercial exploitation of our adolescent side in its relentless restructuring, the permutation of things and brands as novelty, and the curiously intolerant pursuit of fashion.

Further, there is reintrojection of the ideal of transient forms and values into evolutionary settings, so that it can be argued with solemn authority that for man biological evolution has ended and its place has been taken by "cultural evolution." The ideological, stratigraphic mind leaped to this marvelous discovery that social change, transmitted by learned information, has supplanted an objectionable biological process dependent on (the much too slow) genetic transmission.

It was just what society needed to disengage itself from the unsavory determinism of Darwin's thought without having to disprove it. It enabled the humanists, intellectuals, artists, or social scientists to go their autotelic way, untroubled by the deep past, their animal selves, or rules that seemed given rather than made.

Today in the face of the evolutionary studies of culture, the intellectual dimensions of "primitive" thought, the profound biological adaptations to and through which culture is possible, and research on primates, that biological/cultural dichotomy cannot be articulated with a straight face. But society in general lags far behind. The misconception of social "evolution" replacing biological evolution will take many years to fade. The bogus homology of social customs or cultural forms and evolutionary adaptation remains a general feature of the modern intellect. In its own idiom this relativistic monism turns even our humanity and our ecological connections into an ideological choice.

Thus it can be argued that all views on the use of nature are in their way valid—or that all are at least legitimate claims in a democratic society. So decisions will be made on the basis of "interest group" powers. "Environmental mediation" becomes the latest profession to spin off. A key phrase is "trade-off." Economically we have become environmentally hip. We now know we must internalize environmental costs. Thus do we assimilate nature as one more variable.

There remain in the game no givens. The intractable nature of biogeochemical cycles and the requirements of the soil are but temporary obstacles in a world where every element of the ecosystem or the biosphere, like every human, is regarded as having its price. Such an approach is not confined to a blatantly homocentric view of humanity; nor is it merely the spoiler's view (the real estate and building industries). It is the prevailing mode in those government agencies and educational institutions that hire and train the professionals who deal with "land use": agronomists, hydrologists, foresters, and so on.

One could interpret Aldo Leopold's concept, the "ecological conscience," as social expediency: the obligation to conscientiously hear all claims to the control of the natural earth system. But that is not what Leopold meant. In effect, Leopold was rejecting the tenets of his own professionalism. In the preface to his book, *A Sand County Almanac,* he admitted to being one "who cannot live without wild things." Thus he seems at first merely eccentric, but in the end the reader is compelled to ask the same question of himself. To those who are assured that the big questions of our time have to do with peace, race, poverty, economics, and politics, the question may seem frivolous.

But the wild, taken to mean the whole community of species, is the prior question. In fact, it is not a question at all. For there is no alternative to living with wild things, only a world increasingly shaped after the desert regions of the Near and Middle East, where political wrangling and endless marauding are the tattered epilogues to a world goated to death, where people play out their roles like regressed primates, obsessed with the details of the social contract, the final inheritors of five millennia of ideological absolutism and ecological relativism.

It is not my intention, however, to enter into an ideological style in criticizing it: to argue a position and advocate that we relinquish one attitude for another. It seems unlikely to me that one can solve the dilemma by participating in its skirmishes. To insist that we adopt the view that our problems stem from the inconsistency of our views goes nowhere. I would suggest rather that the incapacity of the modern mind to find permanent environmental attachments is a consequence of defective ontogenesis.

The processes of establishing relationships vary enormously over the life span. The relativistic position is both a symptom and preoccupation with adult alienation. It neglects the identity-forming stages of early growth of which it is the defective outcome. Convinced that relatedness to the nonhuman as well as to others is made by the calculating consciousness of the adult, we have abused the needs of the child. Those needs and processes, given the opportunity, lead to a stable concept of selfhood that embraces the earth as firmly as it does the mother.

I cannot elaborate here on these processes, which have been explored recently by Erik Erikson, Harold Searles, James Fernandez, Edith Cobb, and others, but only remark that if the existential dilemmas of chronic relativism are indeed symptoms of inadequate nurturing then we have much reason to hope. For the child is predisposed to enter into the cultus in ways that are irreversible and subsequently undeniable. The total wisdom of humanity may not thereby be increased; its ecological effects will depend on just what the particular vision is. But the development of a mature identity inevitably reaches out to all things, to the growth of an organic relationship in thought as well as fact. In contrast, identity based on arbitrary and transient accretions to the selfhood is a kind of endless smorgasbord in which mature nausea follows adolescent appetite.

 # Sociobiology and Value Systems

THE INTENSITY of the ongoing debate over the significance of sociobiology[1] in our understanding of humankind indicates that the question is part of a larger concern. Its passion and force reveal that it is seen not only as a controversial analysis in a scientific field but as an unresolved inquiry in a larger sphere of human identity. In one sense it is part of the social dialogue that continues to try to resolve the impact of the theory of biological evolution in a culture for which that theory was a painful blow.

It was painful because it struck at the essential duality of the prevailing view of the relationship of man to nature. The evolutionary debate was not a simple matter of theologians against scientists. Thomas Henry Huxley in biology, Pierre Teilhard de Chardin in paleontology, for example, struggled respectively to preserve altruism as a uniquely human domain and to co-opt the purpose of evolution as a river through the human mind and spirit. This obsession with preserving dichotomy also works profoundly below the surface of its logic. It goes to the roots of Occidental thought.

Nothing identifies those who have lived in that ethos better than their rationalization of human alienation. It is as though the ancient Hebrews, whose fate it was to drift between ancient agricultural theocracies and true desert nomads, created an ideal from the conditions of their plight.[2] In the millennia since then, their sense of isolation and disconnection has become deeply inured in assumptions about the world held by their heirs—from sophisticated Greek thinkers and ascetic Christian fathers to contemporary philosophers of disengagement and personal psychological angst.

Westernization is a complex development of those beginnings in Hebrew kerygma and the abstractionism of classical Greek speculation. Those are the deepest roots, but they cannot account for the later forms of cultural rummaging, the war against nature, and the loss of the sacred in daily experience. The historical accounts of religious opposition to evolutionary theory in the nineteenth century have tended to enshadow the intensity and nature of the secular antagonism to it. Renaissance humanism and German philosophical idealism combined to define human experience by contrasting it to that of other animals, to presume an ad hoc priority of difference, and to hypostasize exclusion and duality. With their theories of knowing and perceiving in psychological terms, they gave new life to Platonic values and the arrangement of all things primitive, prehistoric, bodily, and animal in opposition to those civilized, historic, mental, and human. The rhetorical reiteration of these premises, fostered by Italian humanism, was formulated anew in terms of mental criteria. This philosophical antinaturalism now conditions much of modern life— so diffused into the tissues of society as to become a mode of perception.[3] As such it shapes modern intellectual life as well as the practical goals of a technological age.

It is a curious consequence of this history that Christianity, humanism, and technology, although at odds in many respects, are united in their hostility to integrating and mediating natural history and natural philosophy. Whether they derive their inner logic from puritanical Protestantism, from the relativistic fallout of philosophical idealism, or from the Baconian will to subdue and spend the earth, they are one in their formulation of human destiny as largely independent of the natural world.

Literary and academic disciplines including the sciences, in which many of us spend our lives, both give and take shape from these three philosophical focuses of Western culture. Their scorn of animality as an informing center of our most revered human traits is at home in the departmentalized atmosphere of the university, for such contempt is inadvertently fragmenting. The lack of coherence in modern society invites formulations of human and personal identity largely in terms of ideology. In a secular and plural culture the vacuum is filled by a phantasmagoria of autistic explanations. Who and what we are in an ultimately mysterious universe are questions addressed without common ground by disciplines or creeds in the spirit of self-fulfilling definitions. From anthropology's preoccupation with comparative ethnography to zoology's Aristotelian hierarchy, freedom is seen as the right to ontological relativity. Freedom from natural determinism is weakly differentiated from freedom from political tyranny. Ironically, the bleak inadequacy of this free-floating eclecticism is experienced as hostility to the world and further distancing from nature.

Emerging from this truce among Christian-historical-scientific fields, the social sciences in particular are heir to a jealously guarded independence from the givenness of nature. In a society where natural philosophy is equivalent to the history of science and nature is understood in terms of the operations of the physical sciences, it is not surprising that a critic of sociobiology should insist that "socialization is not a concept that can be fitted into physical theory."[4]

In geography, sociology, psychology, political studies, and history, the most salient common feature has been a commitment to the dictum that "Man makes himself." Just as monotheism produces schizophrenic worldviews, this homocentric bias of *Homo sapiens* is the legal tender of duality in the commerce of "cultural evolution," "creativity," and possibility. In the idealization of mind, abstraction, like freedom, is seen as one of the riches of nature; something to be grasped; something that can be held only so long as the antinomies, mind/body, human/nature, genetic/learned, and all the other oppositions are clung to.

Of the three great paradigms of the cosmos—hierarchy, mechanism, and organism—we have perhaps moved slightly toward the organism in the last decade. But that tentative advance is made on thin cultural ice, for the maternal or feminine shape of the ecological vision is foreign to Western presuppositions.[5] The patriarchal view has a vested interest in chaos and meaninglessness, in dread and fear, in linear connections between depths and heights, in distinct levels, in maintaining apartness.[6] Kant, Hobbes, Hegel, and their existentialist and phenomenologist descendants agree that man is not a natural species and neither self-awareness nor the study of human society is to be understood as homological with animals. In this they continue the ancient demythologizing, the obsession with the fallen state, the geometrizing and mechanizing of the universe, and the emerging myths of mastery in the dim rituals of fossil-fuel man.[7]

Every element of society, as an heir to this complicated past, receives its portion of that bias and has its investment in the distancing. The anxieties generated by this fragmenting and disabling rift between people and planet are in turn channeled to expand rather than repair it.[8] The refining definitions of phenomenological ontology, self-centered therapeutic styles, engineering–environmental technique–fix, escapist religious and drug cultism, and the political "solutions" to demographic and ecological imbalances are alike in their reflexive sheltering in the apartheid tradition rather than moving beyond it.

In his book *Charles Darwin: The Naturalist as a Cultural Force,* Paul Sears shows the essential connection of evolutionary theory to the rise of modern ecology.[9] Ecology in turn was the parent of ethology, the framework of sociobiology. Yet it is clear that the main thrust of sociobiology—the homological nature of human and animal behavior, based on the principles of phylogenetic, zoological kinship—is rejected explicitly or implicitly by the dominant institutions and philosophy of the modern West. Attempts to qualify the relative place of the details of an animal or instinctive heritage are, mostly, hedges against the whole of it: compensatory exercises in dialectics. To aver that "man is more than an animal" is only meaningful to those for whom to be an animal is a known quantity, an issue settled by exclusionist principles. If the recent studies of the minds of whales, elephants, chimpanzees, wolves, and other animals reveal that we do not yet know what the boundaries of animality are, then all previous distinctions based on premises of known fields-of-being are invalid.[10] No doubt this new uncertainty only aggravates the unstable definition of self and human deduced from rigid and therefore brittle boundaries.

For men of the age of Arnold Toynbee, civilization is the product of an "evolution" of culture and is embedded in the concept of history.[11] Culture is perceived as superseding evolution and extending its progressive aspect at a greatly accelerated rate. But history is not a record of events; it is an ideological framework exempting (Western) man from the constraints of season, place, nature, and their religious integrations.[12] History is the desacralizing of the world based on writing, prophetic intrusion, and opposition to the natural order. It is precisely not what it seems—the evidence of continuity with the past. It is instead a convulsive break from the true deep past, a divine intercession, full of accident and radical novelty.

This history speaks silently in us. It compels by an irrefutable logic of enculturated alienation. Sociobiology is an anathema that demands opposition by the continuation of dichotomy. Since Descartes we have defined the human by contrast and disjunction, compelling a perception of the nonhuman as deficient and a tenacious resistance to evidence of shared qualities on the ground that it would diminish us.

Like ecology and other "holistic" ideas, sociobiology is intolerable for many—not because its adherents misconstrue the nature of the world but because that nature is unacceptable. It invokes kinship rather than caricature, homology instead of analogy, contiguity rather than parable. The use of animal types as messages remains a respectable literary and linguistic means only so long as all are agreed that human and animal behavior, motivation, mind, and mood are alike only superficially. Otherness must remain outside ourselves. Margins, penumbral similarity, and conceptual umbilicals trigger the schizoid alarm at the heart of the modern Western personality.[13] There is almost no limit to what we will do to avoid that intrusion of otherness into the citadel of prideful identity—including, if need be, exterminating the Others.

NOTES

1. Sociobiology as defined in Edward O. Wilson, *Sociobiology: The New Synthesis* (Cambridge, Mass.: Harvard University Press, 1975).
2. Herbert N. Schneidau, *Sacred Discontent: The Bible and Western Tradition* (Baton Rouge: Louisiana State University Press, 1976).
3. John Black, *The Dominion of Man* (Edinburgh: University Press, 1970).
4. Stuart Hampshire, "The Illusion of Sociobiology," *New York Review of Books,* October 12, 1978, pp. 64–67.
5. Alan Watts, *Nature Man and Woman* (New York: Pantheon, 1958).
6. Mary Daly, *Beyond God the Father: Toward a Philosophy of Woman's Liberation* (Boston: Beacon, 1973).
7. Lewis Mumford, *The Pentagon of Power* (New York: Harcourt Brace, 1964).
8. Theodore Roszak, *Person/Planet: The Creative Disintegration of Industrial Society* (Garden City: Anchor, 1978).
9. Paul B. Sears, *Charles Darwin: The Naturalist as a Cultural Source* (New York: Scribners, 1950).
10. Gordon B. Gallup, Jr. et al., "A Mirror for the Mind of Man, or Will the Chimpanzee Create an Identity Crisis for *Homo sapiens?" Journal of Human Evolution* 6 (1977):303.
11. Arnold Toynbee, *A Study of History* (New York: Oxford University Press, 1947).
12. Schneidau, *Sacred Discontent.*
13. Harold F. Searles, *The Nonhuman Environment* (New York: International University Press, 1960).

Ugly Is Better

"What's wrong with plastic trees? My guess is that there is very little wrong with them. Much more can be done with plastic trees and the like to give most people the feeling that they are experiencing nature."[1]

We may have to make do more and more with "proxy" and "simulated" environments says the author of this statement in *Science*. What is natural and rare is only relative, he adds. And if it is rare, it isn't necessarily worth preserving. After all, who is more important, man or nature?

No doubt he intended to shock the reader, especially those naturalists who sentimentally suppose the natural to be better than the artificial. His is the culminating statement of the recognition of intan-

gible values, the high point to which two centuries of nature aesthetics have led us. He has taken the aesthetes and hoisted them on
the point of their own logic. Given natural beauty as we now understand it, the ersatz is as essential and good as the real. The real solution to the Los Angeles smog problem is to put perfume in gasoline.

The "anti-litter" and "Keep America Beautiful" campaigns were
probably, in all, a worse disaster for the American environment than
the Santa Barbara oil spill. The Spanish, and the Spanish parts of the
New World, were never infected with puritanical tidiness until
recently. Like much of the non-Western world, they accepted the
smells of the body and the reality of excreta as necessary aspects of
life. Anti-litter campaigns and freeway plantings are Airwick and
deodorant soap—sensory crutches protecting our own perceptions
from unwelcome data. In rural Mexico and Spain, wrappings and
bones and junk are still just thrown out—measures of the use of the
world, reminders not just of our consumption of things, but of our
gorging on the ecosystem. Every bread wrapper is a score in our war
with nature that should be seen a thousand times. In fact, it is a double score: first because it wasn't necessary to begin with, and some
tree was cut to make bread wrappers, and second, it simply marks
success in our caloric demand. Pie wrappers and all other luxury
containers signal to us a third score: superfluous consumption.

It is not only that litter is judged ugly in its lack of asepsis. More
importantly, it is not beautiful. It has to do with the category of aesthetic values. Aesthetics is that invention by which sensory qualities
could be dissociated from things and classified abstractly. The term
"landscape" came into use in the sixteenth century to represent the
pictorial abstractions of ecosystems. Such pictures were at first imaginary scenes composed from literary images and were soon formulated by aesthetic theory. Places were in time classified as sublime,
beautiful, picturesque—or without aesthetic significance. It all
became dialectic and esoteric, a proper subject for the leisured, educated connoisseur. Its eventual breakout into the realm of public concern took place in the nineteenth century as part of the spoils system: not as opposed to the spoils system, but as part of it.

American attitudes in the nineteenth century seem ambiguous; there was the common "root, hog, or die" and the great spoiler barons in land, timber, and oil; but there was also Central Park, pastoral graveyards, Yosemite, save the buffalo, Burroughs, Muir, the Audubon Society, and an enormous popular addiction to picturebooks and sentimental nature poetry in every newspaper. It looks at first like counterculture, and it may have been for some. Mostly it was the system taking over the old landscape aesthetic, one with which it could live, and making illusory options—like the modern soap company that in reality owns its own competition.

Look what the industrial society could do with landscape aesthetics. It could shunt it into pictures and other symbolic tokens; it could be geared to style, taste, and fashion in that order—the clear-cut slope and strip-mine spoils really are beautiful, they just don't happen to be a la mode; the beautiful places could be identified and isolated from the rest of the biosphere. And the qualities could be de-totalized and translated into technique: A patina of pastoral planting was laid on virtually every college campus built after 1850.

And the concern for scenery was profitable. Somebody said to me during a trip to the desert, "This is great. I can't wait to get home and look at my slides." To appreciate what it has meant to the travel industry, you must travel in a place where there are no accommodations for tourists. But more important, scenery unfettered the spoilers; nature is really resources—except where we have made parks—because man does not live by bread alone. Aldous Huxley once observed that Wordsworth was inapplicable to a tropical jungle. But he needn't have gone that far. Wordsworth doesn't apply to much of Texas, Georgia, Alberta, Baja, or New Jersey.

Conservation is the rubric under which landscape aesthetics was incorporated into enlightened exploitation. Officially it had to do with spiritual values, but for the hard core it could always be translated into money values by feeding its raw data of participation through a translating machine called Recreation. "Scenic Resources" fit well with "human resources." Then, for the corporate agencies,

1970 was the traumatic year of confrontation. The mountain heaved and gave birth to two peas: two changes in terminology. They struck out "conservation" and "nature," inserting in their places "ecology" and "environment." A great rhetorical year.

The difficulty is that it is practically impossible to discuss our experience of the nonhuman without recourse to a jargon that is the property of an outmoded and destructive enterprise. Worse yet, in the field of action it is the same. Recycling is the ecological slave in the front office. We seem determined to engage in the most frenetic charades and games to avoid reducing consumption and human numbers. The strategy of the system and the options provided by the barons have always been to quietly provide harmless alternatives.

To hell with conservation and nature aesthetics. The confrontation with the nonhuman occurs every second. Every breath is an encounter with nature and every bite of food is part of a language. We have been conditioned to reserve feeling and thoughtfulness and attention to the nonhuman for our visits to those scenic enclaves or their pictorial representations. For whooping cranes and rhinoceroses we may indeed have to provide protected terrain, but that is the last desperate measure, not the best one. The protected lands on which threatened species live should not be open to the public at all: The species should be regarded as in a retreat from which they may once again emerge in a functional relationship with people. To reduce creatures to spectacle is part of the game, making them merely beautiful. The famous remark, "Seen one redwood, seen 'em all," is true. It refers to retinal forms, curiosities, architectural and pillared spaces through which one moves; they are objects from another world, repetitive surfaces filling the visual screen. The observation has candor and courage; it puts the aesthetics of the beautiful where it belongs.

The nongame alternative is that the redwoods are beings. Since they are more remote from us than other human cultures and races, more circumspection, not less, is necessary. We cannot so easily "know" them. If there are means of doing so, they are long neglected by our culture. If there are no means of doing so, then the mystery

itself is manifest. Perhaps both are true. In any case, we cannot for-
mulate a new relationship out of air. Religiosity is the trap that idealism and ideology set for the antinomian. We cannot achieve a fundamentally different worldview by an act of will alone—some
individuals can, perhaps, but not societies.

For the present it is just as well. We have only begun to recognize the extent to which the Faustian hubris has usurped aesthetic
and ethical categories. We have just recently started to appreciate the
modes of consciousness possible and to apprehend the incredible
richness and otherness of nonhuman being and the impossibility of
surviving a man-made world. A century of ethology has hardly
touched the ways of being open to other species and the ecological
wisdom that has been realized some places and some times.

This is not a cop-out. It is not the curiosity of the inventor and
capability of the engineer that have been at fault—but rather the zeal
to employ every technological innovation for change and newness as
ends in themselves. Changing culture is open to the same mistake. It
is not simply that action must be preceded by understanding; it is that
at present further understanding is the most important action.

If you must have some symbolic actions, I recommend the following: Throw your wrappers, papers, butts anywhere, beer cans in
the streets, bottles on the berms and terraces; uproot and cut down
all ornamental trees—replace them with native fruit-bearing trees
and bushes; sabotage all watering systems on all lawns everywhere;
pile leaves, manure, and garbage among growing things; return used
oil, tires, mattresses, bedsprings, machines, appliances, boxes, foil, plastic containers, rubber goods, and all other debris to its origin—seller
or manufacturer, whichever is easier—and dump it there; unwrap
packages in the place of purchase and leave the wrappings.

When this has gone on long enough, some tokens of the glut of
overconsumption will at least be evident. Equally important, there
will be less refuge from the countryside with its regimented mono-
cultures, scalped slopes, poisoned rivers, and degraded rangelands.
Our society goes for letting it all hang out, so let's do it. Are
encounter groups in? Let's raise the encounter a whole octave and

confront the real human ecosystem that we live in. Some great Avon
Lady keeps rouge on the cheeks of the middle-class neighborhood,
the industrial park, and the college campus; the same tinsel earth
mother in whose name the slaughterhouse is hidden, the zoo's dead
are unobtrusively replaced, and the human dead are pseudo-fos-
silized.

We may, as the Sierra Club maintains, need wilderness as a spiritual tonic. But if so, it has nothing to do with glorious picturebooks
or even with landscapes. For John Muir, the club's founder, the landscape was the canvas painted by God. Henry Thoreau, by contrast,
knew better. He edited no picturebooks, did no landscaping. Looking for kindred spirits, he once read William Gilpin, the English vicar
who also wandered over the hills. Gilpin observed that a horse was
aesthetic because of the effects of light and color in its coat. Thoreau
said: "And this is the reason why a pampered stud can be painted!
Mark that there is not the slightest reference to the fact that the surface with its lights and shades belongs to a horse and not to a bag of
wind." The observations of terrain in Thoreau are prospects, the
descriptive opposite of the landscape scene.

❧

THE PROSPECT was the unfettered view of the ambient from a high
place. You can see it in the paintings by the elder Brueghel. The
prospect is not a pastoral dream or an interesting texture. In one of
the few paintings of a man cultivating the soil (*The Fall of Icarus*)
Brueghel has included the unscenic details, and Icarus splashes in the
distance while the horse goes on farting down the furrows.

Despite the old masters' perspicacity, pictures themselves are part
of our present problem. Theirs was an iconic reality and Brueghel
could not have foreseen our dilemma. He was not concerned with
the picturesque ("the scenery's capabilities of being formed into pictures") as Christopher Hussey has described it in *The Picturesque* or
the terrain's "capabilities" for being reshaped into garden landscapes
in imitation of old paintings by eighteenth-century landscape architects like Capability Brown. The substitution of pictures for places

was the step toward making places that match pictures. Now we are taking pictures of places whose patterns happen to suggest those gardens built in imitation of paintings which were originally done as visual expressions of literary evocations of "classical" scenes.

Scenery is from a Greek word meaning stage props.

NOTE

1. Martin H. Krieger, "What's Wrong with Plastic Trees?" *Science* 179 (1973):446–454.

 # Itinerant
Thoughts
on Place

I T IS NOT SURPRISING that the individual developmental processes
which comprise that sense of self often called "identity formation"
are generally regarded by modern scholars as given wholly by the
human part of the environment: "Nature" is that tiny human animal
rescued and shaped by "culture." Not surprising because of the
mockery made of the nonhuman by turning it into scenery, the way
picturesqueness reduces everything it touches to surfaces. From the
moment the grand tourists invented modern tourism, the gentility
went about Italy speaking of the genius loci as though it were a landscape painting. Then it was hitched to consumer recreation. The admir-
ers of landscape were never the opponents of the Promethean hubris,
only the disguised and sometimes unconscious handmaidens of it.

The old Romans, whose poetry inspired the notion of genius loci, did not mean pictures at all but a tutelary divinity, a guardian spirit. It was the same among the Greeks, whose temples were expressions of the character of particular goddesses in whose laps the structures were placed. The subtlety with which their architecture was accommodated to the terrain included even the configuration of the horizon; the temple passages were designed to guide ritual processions whose central themes were a dialogue between the people and the earth.

However much they admired the old arts, the neoclassicists more than a millennium later could feel little of the old pagan interior sense in which these sacred places were experienced as part of themselves. The Jews and Christians had methodically sought out the old shrines and covered them with churches in order to redirect the religious impulse away from that place. The bishops who consecrated them and the liturgy they followed referred to a Holy Land elsewhere and a heaven and hell that were nowhere and everywhere.

In his widely read book *The Sacred and the Profane,* Mircea Eliade has instructed a whole generation on the making of sacred places: the rites and ceremonies that "cosmosize" a hearth or an altar. But there is for him no real chthonic, no real spirits, only human beliefs. Although at pains to insist on the religious man's loyalty to the heterogeneity of space, he sees it only as something made by men. The indigenous qualities of the spring or cave or mountain are for him little more than markers. There is not the slightest hint that the spiritual entities which preclassical Romans, Greeks, or other so-called primitives conceived as indwelling were anything but cultural assertions. One called in the forces from a centralized heaven the way one dials a long-distance operator.

Of course the Christians did not invent this making of place by will and designation. The ancient civilizations of the Middle East are speckled with temples built where they would be convenient to the bureaucracy, the keepers of the grain, and the army barracks. The shift of attention away from the uniqueness of habitat began long before the Church fathers declared that all places on this earth are pretty much the same. Eliade's view is ultimately no different from that of municipal street-namers: The world behind the human facade is homogeneous. One finds place rather than discovers place. It doesn't matter whether a priest blesses an altar or the mayor cuts a ribbon. The autochthonous forces by which the earth speaks are only elements in myths by which the peasants rationalize their designations of sacredness. Quality is not given, it is made.

But the polarity and mystery of the given and the made will not go away. It is the duality at the heart of knowledge, the central enigma of our private and collective identities. Arthur Modell has observed that the painting and sculpturing of the Paleolithic caves of Southern Europe often used the erosional forms of the rock as the basis of the animal figure. This synthesis of what is there and what is created externalizes an inextricable relationship between the artist and his materials, between ourselves and our bodies, man and nature. The artist affirms and formalizes that tension which is the core of the maturing self-consciousness of everyone. Art, says Modell, is always a love affair with the world. It was this search for a self that was not solely defined by acts of transcendence and domination which moved the nineteenth century's romantic imagination, feelings for which the Augustinian humanist and modern highway/pipeline/parking-lot builder cannot conceal their scorn.

From nutritional and environmental studies of laboratory maze-running rats to observation of babies with and without playpens, institutionalized children, and the psychology of the playground, the evidence is strong that heterogeneity is like an essential nutrient. But how does it work? You cannot, after all, just put a baby in a bag with a thousand objects and shake well before using. Claude Lévi-Strauss believes that the species system of plants and animals is a durable, dependable concrete model for the development of the powers of categorizing—that is, basic cognition. Edith Cobb holds that the fixation on terrain is an organizing process by which the percept of relatedness is interiorized. White and his associates at Harvard find that the intelligence of children emerges relative to a spatial movement among objects, coupled with naming. All of these imply real

changes in the nervous system. Psychiatrist Mayer Spivak has identified thirteen "archtypal places." One can live without these special
places for resting, feeding, conviviality, grooming, courting, and so on.
But without them we become, as do other similarly deprived mammals, increasingly stressed and pathological. Such places are not
merely spaces for different activities but are the perceptual and physical prerequisites of different behavior.

The habitat is not merely a container but a structured surround
in which the developing individual makes tenacious affiliations.
Something extremely important transpires between the complex
structure of these particular places and the emerging, maturing self.
What is going on is a kind of macro/micro correlation—a centering
mostly unconscious but essential to the growth of personal identity.

✷

EVEN IN ITS MORE CONVENTIONAL SENSE—identity as our conscious position in matters—place is important. Before the revolution
the American knew himself in three contexts: as Protestant, English
colonial, and village community member. As this scaffolding was cut
away by revolution, secularization, and industrial-urbanization,
Americans of the nineteenth century suffered an acute attack of
inchoateness from which they still have not recovered. The landscapes
of those institutions had been the stable rural countryside tightly and
hierarchically ordered around the church and town, making zones
upon the land and interpenetrations with the wild that changed little between 1520 and 1800. In an era of rapid change we may forget
how constant was American life for three centuries and overlook
how traumatic was the collapse of that old order. In the paintings of
the Hudson River school we have a body of representations of terrain preceded by an indigenous prose and poetry, especially that of
Washington Irving, J. F. Cooper, and William Cullen Bryant. From
time immemorial the myths of creation have been presented as
drama, perhaps could only be comprehended that way. The "legends"
of Sleepy Hollow and the adventures of Leatherstocking among the
Iroquois were geographically explicit. It was possible to go, as John

Trumbull did in 1810, to Norwich Falls in Connecticut, where Cooper had placed a climactic scene in *The Last of the Mohicans,* and do a portrait of the place—an analysis of its character the way one would a person. Trumbull was a painter of heroes and battles whose work hangs in the Capitol in Washington. He probably painted no more than three landscapes in a busy career. The fictional heroes and events, no less than the historical, gave place to elements of American identity—for we "identified" with the Irving and Cooper characters.

From the "actual" sites of such places, which much of the early painting scrutinized with an almost frantic intensity, the painters moved out in search of correspondingly dramatic sites, appropriate to the imagination of episodes of pioneering life or storms or mountain geology. Thomas Cole was its most fervent spokesman. He painted and wrote long essays. To be lost in the wilderness, he said, was the supreme experience. Neither the eroticism nor the adolescent emotionality of his work have gone unnoticed. This immaturity (for which critics later had only contempt) had its purposes, however, for it turned Americans back to maternal themes—to the land in a search for new beginnings without which they would remain lost. Cynics after the Civil War saw the romantic artists merely as weak and undisciplined, as the Victorians considered women to be. To attend to trees and waterfalls seemed to them sentimental and silly. The artists' own personalities were indeed rife with immaturity. But they were, in a sense, childish for us all.

Before them, the Europeans who settled America were on alien ground. With few exceptions their concrete connection to locality remained in Europe. The painters in America tried, in a few decades, to overhaul that whole troubled subjectivity, to imprint on our nervous systems a new Eden in the form of a wild mountain spectacle as home, to do in the New World what had taken centuries in Europe. Of course they could not possibly succeed. But in some ways no concept of place and landscape in America since then has been without something of their mark.

We are reminded with painful regularity of our continuing sense of dislocation, the neuroses of personal identity problems, the terror of loneliness in the crowd, of isolation both from society and from the rest of nature. These anxieties are fraught with doubts about the purpose of life, even of order in the creation. Traditional psychology, peremptory humanism, and even our religious preoccupation with the self have tried to explain these dilemmas of disconnectedness as arising within society and its works: in the family, the home, the job, or the church. But our failure to either elucidate or remedy them on those grounds raises doubts that man either makes or unmakes himself apart from the nonhuman, or that his loneliness stems from inadequate social planning or ideology.

It is easy to blame rootlessness, mobility, the fluidity of American life for our anguish. But all the truly primitive people—the hunting and gathering cultures—who have ever been studied move serenely through hundreds of miles without such troubles. Although they traveled through vast spaces there is a scale about their lives that is different from ours. W. H. Auden once observed that the mega-world of galaxy and the mini-world of the atom are real for us mainly in frightening ways. He wrote of a middle world. "Saving our sanity" in this mesocosm might well include putting the heaven of the other world religions along with the Adlerian psychology of simply willing your own world with other tall stories.

Eliade wants us to believe that places differ according to the amount of a universal holy oil we pour on them. He is in company with the cartographers whose surveys of latitude, longitude, township, and range we have also accepted as the terms of location and "defining space." But the world is not a billiard table until we finally turn it into one. It is unique everywhere in combining features differently that, in some unknown way, both reflect and create an inner geography by which we locate the self.

However exact the mathematical, political, or ecclesiastical subdivision of space may be, if it is imposed from the outside it cannot refer to place in the sense that is meant here—any more than maturity is achieved by ceremonies by which those institutions confer power on the individual, however much ceremonial scenery they frame it in. Tyrone Guthrie, the British director and drama critic, once wrote of Thornton Wilder's play, *Our Town:* "Wilder uses the stage not to imitate nature, but to evoke, with the utmost economy of means, a series of images. . . . It is one of the paradoxes of art that a work can only be universal if it is rooted in a part of its creator which is most privately and particularly himself. Such roots must sprout not only from the people but also the places which have meant most to him in his most impressionable years."[1]

Wilder may seem for us to create place—or, more exactly, he creates the means of its recognition. But his own experience is one of discovery. He is like Carlos Castaneda, who tells us how difficult it is for one coming from a culture of the human domination of nature to discover, even in a room twelve by eight feet, the spot on which he could sit without fatigue. His frustrating search under the tutelage of Don Juan took all night. Again one thinks of the nineteenth-century painters on foot, roving back and forth across the White and Green mountains, the Berkshires, the Taconics, the Hudson Highlands, and the Catskills searching endlessly. You don't need to connect with Castaneda's brand of spirituality to realize that the heterogeneity of the land is not made by humans—only discovered and celebrated, or ignored and diminished, by them.

One is reminded of John Ruskin's refusal to come to America, saying that he could never visit a country which had no old castles. If you had no old castles you had no history. And if you had no history there was no place in which the events, which made you, sanctified the ground.

But Ruskin was inordinately attached to the picturesque, to the necessity of ruins, and to the moral qualities of painting. Some dimensions of place do not depend on architecture. In her introduction to the poetry of Carl Sandburg, Rebecca West described the loquaciousness of Americans in public, their readiness to discuss their lives with total strangers, and the leisure they take in self-explanation. She says: "It occurs to one, as such experiences accumulate, that one has encountered in art, though not in life, people who talk and behave like this: in Russian novels. There one gets precisely the same universal addiction to self-analysis. And then it occurs to one also that this place is in certain respects very like Russia . . . is a high spot, to

use the idiom, on the monotony of the great plains. . . . It profoundly
affects the language. . . . In the Middle West more than anywhere else
the introspective inhabitants have developed an idiom suited for
describing the events of the inner life and entirely inadequate in dealing with the events of the outer life."[2]

Pictures and parks are part of that American syndrome. Tourism
and the park mentality, like that which pushed American Indians
onto reservations, make enclaves—not on the theory that quality is
everywhere unique but on the theory that between the isolated
points of interest there is only weary uniformity. Carl Sandburg was
not singing parks and petrified villages like Lincoln's New Salem. He
belonged. Belonging, says Erik Erikson, is the pivot of life: the point
at which selfhood becomes possible—not just belonging in general,
but in particular. One belongs to a universe of order and purpose that
must initially be realized as a particular society in a natural community of certain species in a terrain of unique geology. What Rebecca
West sees as the empty plains of Illinois and Russia betrays her own
bias. For they are empty only—as she noted—of close neighbors.

✂ⵣ✂

MY THEME CAN NOW BE DRAWN TOGETHER and signified by the
"walkabout" of certain Australian aborigines. In going on the pilgrimage called walkabout, the aborigine travels to a succession of
named places, each familiar from childhood and each the place of
some episode in the story of creation. The sacred qualities of each is
heightened by symbolic artforms and religious relics. The journey is
into the interior in every sense, as myth is the dramatic externaliza-
tion of the events of an inner history. To those on walkabout these
places are profoundly moving. The landscape is a kind of archive
where the individual moves simultaneously through his personal and
tribal past, renewing contact with crucial points, a journey into time
and space refreshing the meaning of his own being.

Terrain structure is the model for the patterns of cognition. As
children we internalize its order as we practice "going" from thought
to thought and learn to recognize perceptions and ideas as details in

the sweep of larger generalizations. We intuit these incorporations and textures into a personal uniqueness. Mind has the pattern of place predicated upon it, and we describe its excursions, like this essay, as a ramble between "points," the exploration of "fields," following "paths," and finding "boundaries," "wastelands," or "jungles"—of the difficulty of seeing forests for the trees, of making mountains of molehills, of the dark and light sides, of going downhill or uphill. Cognition, personality, creativity, maturity—all are in some way tied to particular gestalts of space: to locality, partly given, partly found.

What does this say to us as Americans? From the standpoint of society as a whole, our disadvantages seem obvious and enormous. We have little cultural continuity with the land; history has few tangible relics. The vestiges of precolonial art and earthworks remain, but their meanings we do not know or feel. At the time of its settlement by Europeans, the continent had vast diversity, as indicated by the various Indian tribes. Almost everything we have done to it in the last century has worked toward the destruction of these differences. We have idealized this uniformity in the image of the melting pot and the standard of living. The industrial complex levels mountains, drains swamps, opens forests, plants trees in grasslands, and domesticates and exterminates the wild. We have long been aware of this—and of the rejoinder that it is a small price to pay for convenience, security, and comfort; that entertainment, travel, instant news, electrified homes, and an unlimited array of goods are made available in this way. Diversity, in fact, is suspect because it is divisive, or at best it is just one more source of pleasure in a complex of "trade-offs." We are doubtful and ambivalent about diversity.

But the lament is not that as a nation we lose the multifold character of a continent or lack architecture that affirms these qualities. We do not experience anything as a people, only as individuals. To the corporation or bureaucracy the quality of place is merely an amenity because our mythology of collective power and metaphor of collective experience confirm the transcendence of the individual.

Children must have a residential opportunity to soak in a place, the same place to which adolescents and adults can return to ponder and integrate the visible extensions of their own personalities. Place in human genesis has this episodic quality. Knowing who you are is a quest across the first forty years of life. Knowing who you are is impossible without knowing where you are. But it cannot be learned in a single stroke. What makes the commercial ravagement of the American countryside so tragic is not that it is changed and modernized, but that the tempo of alteration so outstrips the rhythms of individual human life and the patterns of going and returning.

Everywhere in America we continue to be enmeshed in immaturity, caught in a drama of infantile separation anxieties. Samuel Beckett, in his plays, has rightly set our quandary in an empty landscape, surrounded by a terrain that is both featureless and meaningless, where we wait at crossroads marked only by signposts for something to happen. Signposts do not make a whereness nor beliefs a wholeness.

If we were all as alike as eggs, it would not matter. But we live in a world where everything we do can make or unmake the possibilities for our further growth. We academic eggheads like to think that we live in a world of ideas which we invent—that we create the domestic plants and animals. But in some part of our skulls there is a wilderness. We call it the unconscious because we cannot cultivate it the way we do a field of grain or a field of thought. In it forces as enduring as climate and bedrock maintain our uniqueness in spite of the works of progress. It is the source of our private diversity. Among us, our collective unconscious seems almost to exist apart from ourselves, like a great wild region where we can get in touch with the sources of life. It is a retreat where we wait out the movers and builders, who scramble to continually revamp our surroundings in search of a solution to a problem which is a result of their own activity.

Notes

1. Tyrone Guthrie, *New York Times,* November 27, 1955, p. 27.
2. Rebecca West, ed., *Carl Sandburg's Selected Poems* (London: Harcourt Brace, 1926).

Place and the Child

TWO APPROACHES to the understanding of place are expressions of a larger split that penetrates scholarship as well as public life. The majority position—largely self-formulated and consciously shaped—is that uniqueness is grasped and identity at least chosen from the anomalous streams of social and political life. Much current writing assumes autonomy of place and self. The earth, like the body, is a nascent expanse of no-place until a kind of cartography has posted it. The ideal is to escape deterministic aspects in favor of pure individual creativity.

The minority (and more interesting) position is that there are compelling local forces in the earth itself—forces creating peculiar fields of energy to which people are more or less sensitive and which are typically discovered rather than constructed. Obversely, there are intrinsic, structured, chronological analogs in the individual human being: identity coming from within instead of superimposed or fabricated. The individual is shaped by the constraining given of place as well as by the natural history and genetic constitution of the species. This contrary view is prior to history and has been generally accorded a grudging place in the majority worldview of relativistic and existential values. In 1973 Mayer Spivak called attention to a facet of animal ethology often unnoticed—that large, mobile mammals use many different habitats at certain times of the season or life cycle. To be ecologically niche-bound, as all living things are, is not always to be habitat-restricted. This does not imply a liberation from natural context but, on the contrary, involves a commitment to diversity and adaptability.

It is an appropriate observation point from which to consider human ontogeny—the development of the individual that includes neoteny (the retardation of development and extended immaturity) as well as biological specializations of the highest order. Childhood has been misunderstood in the humanistic literature as a kind of "freedom" to erect a self-world in a cultural milieu that might be characterized as autistic (by my dictionary "a form of schizophrenia typified by acting out"). No wonder there has been historical disagreement as to whether there even was a childhood in the past.

Ontogeny is, succinctly, a program, genetically coded, whose successive phases are keyed to environmental and social signals. The nonhuman coplayers in this maturational drama include an array of animals—the perception and identification of which provide both the cognitive instruments of categorical intelligence and fragments of the inchoate self projected onto their respective stereotyped behavior spread across the landscape.

As a Siberian myth tells it, all the body organs once lived independently as creatures. Something of the sort also keys the child's emotional states, feelings, and behavior to the concrete forms of nature—an external reality swallowed by introjection and internalized to make a self. It is a kind of shadow play of the sacramental rite of eating the god's body to attain grace. The animal infinitives—to duck, to bear, to grouse, to bug—constitute a mosaic of the living repertoire gathered in a finely tuned, unconscious, mimetic, critical period of attention, speech, and nurturance. Ecologically animals are inseparable from place—except as we see them domestically blunted, insane in zoos, or petted as meek surrogates.

Beyond the animals proper there are two more explicit processes in which a copula occurs between the individual and a genius loci. The first process is the imprinting of home range—the terrain where children play and wander and explore—upon their psyches. The second is the response of the community (culture) to adolescent transition into adulthood—a process that in tribal cultures takes the form of initiation rites bound closely to place. In her *Ecology of Imagination in Childhood*,[1] Edith Cobb explains the importance of childhood terrain. Surveying the lives of geniuses, she noticed a common thread: the return in moments of creative meditation to the place of childhood in imagination or sometimes physically—a journey needed to find solutions to life's problems. The original meaning of the term "genius loci" referred to a unique sacred power situated in place. What is it, Cobb asks, about the original experiences of childhood that, revisited and recapitulated, make them useful to the psyche as well as an organizing force in adulthood?

The child, she says, seeks to make a world the way the world is made. Her biographical studies led her to the conclusion that the terrain itself provided the durable gestalt upon which the intellect germinated. "Home range" for the seven to ten year old is the prime, patterned, concrete reality of life, upon which wavering and nubile powers of memory and logic cling and develop, like seals climbing onto the rocks to give birth.

Cobb concludes that adult faith and intuition and order permeate the cosmos, that no bit of datum or bizarre idea is truly disparate, that searching will be rewarded, and that all this extends from the singular imprint of an intensely inhabited space of about thirty-five acres at a crucial time in childhood. Played through, the child's transit, time and again, locks this literal, objective reality into an unforgettable reference for later retrieval. The path traversed by the child becomes a holding ground: a tiny universe envisioned by details of a landscape. The home range of rabbits, culverts, banks, and fences provides a boundary-marking function for half-formed and elusive ideas and relationships. Edith Cobb's own genius has given us insight into the primordial meaning of coherence as a function of a specific, tangible ecology swallowed by the young child in repeated excursions.

The means and implications of this order-making process remain to be understood. We have yet to learn how adventures and play in streams, in tree houses, and along pathways and encounters with mean dogs and inclement weather can lead into those ultimate conjugations of sacred forms, symbols, and ritual acts in later life. We can surmise that early experiences are the fellow players in moments of acute awareness which occupy the space that one of my students calls "the chipped tooth, dead cat landscape." Does pretending to be a machine instead of a tiger make any difference? What counterpart is possible in the city's fabric to the flowing lines of the countryside or the otherness of wild places?

The eventual linking of the natural matrix to sacred forms and symbols is set into play by another place-bound episode in ontogeny. This is initiation: the setting of adolescent transformation, a culturally mediated passage whose understructure is puberty. Its religious connotations have been dealt with at length by Mircea Eliade and others; in another plane, its psychophysical signs and pathologies constitute a whole subfield of psychiatry. That the pubertal child has a precipitate mentality and emotional and intellectual prevision of future unknown events and trials is a culminating phenomenon of childhood. A formal tribal initiation takes this into account. The youth is ushered into adult status by ceremonies that include separation from family, instruction by elders, tests of endurance and pain, trials of solitude, visions, dreams, and rituals of rebirth. There is reason to think that modern anomie and disaffection may have roots in society's failure to respond to this time-critical need in our youth.

Central to the imagery of adolescent passage is the creation of an ideal world—a transitional stage from a literal to a figurative place, from familiar temporal reality to dreamtime. One of its instruments is, to use the psychiatric jargon, the "transitional object." Unlike the small child's doll or blanket, it now refers to places and objects of special significance, stories told, and the enacted metaphors of rituals of death and rebirth in which a leap is made from the natural world to the cosmos.

The transitional objects of adolescence belong to the category of objets trouvés: art found in nature. The most enduring examples in Western human history are probably the Paleolithic caves of Europe. The primordial setting is the cave, and the mime of rebirthing is usually preceded by a journey through a terrain of hills, springs, and special trees given a second meaning in myth. Although I described this journey as a "leap" from one world to another, that term is not exactly right, even though dancing is important to it. Eliade may have been wrong in postulating the exclusive distinction between sacred and profane worlds—an opposition that fits mainly peasant Christian conceptions or other Indo-Aryan dualisms. The older function of the leap is not from this world to the other, but movement like the shuttle in a loom, binding and informing rather than departing. Typically the ceremonial enactment attains identity with a mythical past, not by transcending the physical reality, but by extending its meaning and obliterating the time constraints that characterize historical societies.

The typical adolescent preoccupation with her or his own body is, from this perspective, crucial, for the introjection of space that constructs the body in the child ceases to be the palpable place it was. The body is to be reconstructed along with the world. The special places of adolescent initiation therefore embody a story, simultaneously altering the individual's interior landscape. Place is first conceptualized as a world with named things and relations—the tacit equivalents of an inner constellation—imprinted as the first model of order. Literal places in childhood become metaphorical in adolescence, a syntax for the description of an ideal world.

To summarize: I have mentioned moments of correspondence between inner and outer worlds in which specific features of the earth and its fauna provide children and youth with sensory figures corresponding to latent elements of the self. Body and space are construed in tandem. Closure occurs. Specific places and creatures become unique to the self, obtained on a temporal wind.

The juvenile terrain and fauna and the adolescent map of environmental apotheosis are bandings. The first refers to a concrete matrix encountered in the protorituals of play. The second is a triple matrix of spatial counterparts: the body, the terrain, and the cosmos. In the world of forming identity by superposing a hierarchy (corporation, religion) or by self-proclaimed ideology (communism, libertarianism, feminism)—that is, the entrepreneurial world of making place by free seizure—we come finally and paradoxically to live alienated lives, fostered by myths celebrating arrogance, power, and acquisition. But other myths are possible: myths of keeping and caring and myths of freedom that affirms limitations. We are free, culturally speaking, to internalize place in any way we wish. But we are not free to control the consequences.

What happens to the child who is deprived of Nature? In a cement or desertified home range without creatures, what is the ontogenetic outcome? A barren childhood, I presume, proceeds to an inorganic substrate without the coherence of dense uniqueness. Such individuals will instead project infantile concerns on the empty stage of the world and cosmosize that placelessness as a battleground in which fantastic ideologies struggle for power in the mode of juvenile heroics. Such a world will lack both tolerance of otherness and recognition of the ecological fabric bringing meaning to space, time, and matter—living and nonliving, human and nonhuman. Madness is the kindest word for such a world.

In terms of the "American experience," some of these constituents of an identity take on distinct qualities. Among these are geographic mobility of families, experience of animal diversity, and various efforts to meet adolescent needs. Native Americans have ways of meeting the psycho-ecological needs of children and youth that are available to Caucasian-Americans and can be modeled. Caucasian-Americans, paradoxically, are closer to and farther from indigenous fauna and landscapes than their European cousins. Their lack of traditional identifying events that conserve socially shaping activities not only makes for a deracinated condition but impels society to explore new means of grounding identity—and, in some cases, of recovering modes of shaping identity that our Pleistocene bodies still expect. Despite the rapacious work of the nineteenth-century lumber barons and the twentieth-century corporate energy moguls in homogenizing the continent, much wildness remains. And the fluidity of society is itself perhaps a doorway to the realization of Roxy Gordon's statement that "real revolution is born from genetic memories of ancient reality."

However arbitrary the details of play or the content of myths of origin or the choreography of ceremony, they constitute the inescapable drama of the unfolding personality bound to the necessity of a diverse and healthy surrounding. Maturity is strangely linked to the nature of hatching places, and its outcome is a measure of the resonance of inner and outer landscapes.

NOTE

1. Edith Cobb, *The Ecology of Imagination in Childhood* (New York: Columbia University Press, 1977).

Works by Paul Shepard

This bibliography contains only published work by Paul Shepard in the primary areas of his research: nature perception, human ecology, the hunter/gatherer legacy, human/animal connections, the place of nature in human development, and the bear in mythology and culture. Before his death in 1996 he had completed manuscripts for *Encounters with Nature* (1999) and *Coming Home to the Pleistocene* (1998), which were edited by Florence R. Shepard and published posthumously by Island Press. Book reviews, newspaper and newsletter articles, lectures, and unpublished essays are not listed here. Neither are published essays pertaining to conservation written in the 1940s and early 1950s when he was field secretary for the Missouri Conservation Federation, conservation chairman of the National Council of State Garden Clubs of America, and a member of the Yale Conservation Club. His first book was the history of his battalion, *The Pictorial History of the 493 Armored Field Artillery Battalion, 1943–1946* (Augsburg, Germany: E. Kieser, 1945), which he edited and published after the war when he served as a historical technician with the army of occupation.

Books

Encounters with Nature: Essays by Paul Shepard. Edited by Florence R. Shepard with an introduction by David L. Petersen. Washington, D.C: Island Press/Shearwater Books, 1999.

Coming Home to the Pleistocene. Edited by Florence R. Shepard. Washington, D.C.: Island Press/Shearwater Books, 1998.

The Tender Carnivore and the Sacred Game. With a foreword by George Sessions. Athens: University of Georgia Press, 1998. First published in 1973.

The Only World We've Got: A Paul Shepard Reader. San Francisco: Sierra Club Books, 1996.

The Others: How Animals Made Us Human. Washington, D.C.: Island Press/Shearwater Books, 1996.

Traces of an Omnivore. With an introduction by Jack Turner. Washington, D.C.: Island Press/Shearwater Books, 1996.

Man in the Landscape: An Historic View of the Esthetics of Nature. College Station: Texas A&M University Press, 1991. First published in 1967.

The Sacred Paw: The Bear in Nature, Myth, and Literature. With Barry Sanders. New York: Penguin, 1992. First published in 1985.

Nature and Madness, With a foreword by C. L.Rawlins. Athens: University of Georgia Press, 1998. First published in 1982.

Thinking Animals: Animals and the Development of Human Intelligence. With a foreword by Max Oelschlaeger. Athens: University of Georgia Press, 1998. First published in 1978.

Environmental: Essays on the Planet as Home. With Daniel McKinley. Boston: Houghton-Mifflin, 1971.

English Reaction to the New Zealand Landscape Before 1850. Pacific Viewpoint Monograph 4. Wellington: Victoria University of Wellington, 1969.

The Subversive Science: Essays Toward an Ecology of Man. With Daniel McKinley. Boston: Houghton-Mifflin, 1969.

Parts of Books

"Paul Shepard" (interview). In *Listening to the Land: Conversations About Nature, Culture, and Eros,* ed. Derrick Jensen. San Francisco: Sierra Club Books, 1995.

"Nature and Madness." In *Ecopsychology,* eds. Theodore Roszak, Mary E. Gomes, and Allen D. Kanner. San Francisco: Sierra Club Books, 1995.

"Virtually Hunting Reality in the Forests of Simulacra." In *Reinventing Nature? Responses to Postmodern Deconstruction,* eds. Michael E. Soulé and Gary Lease. Washington, D.C.: Island Press, 1995.

"The Unreturning Arrow" (interview). In *Talking on the Water, Conversations About Nature and Creativity,* ed. Jonathan White. San Francisco: Sierra Club Books, 1994.

"On Animal Friends." In *The Biophilia Hypothesis,* eds. Stephen R. Kellert and Edward O. Wilson. Washington, D.C.: Island Press, 1993.

"A Post-Historic Primitivism." In *The Wilderness Condition: Essays on Environment and Civilization,* ed. Max Oelschlaeger. San Francisco: Sierra Club Books, 1992.

"From Nature and Madness." In *Learning to Listen to the Land,* ed. Bill Willers. Washington, D.C.: Island Press, 1991.

"Objets Trouvés." In *The Meaning of Gardens: Idea, Place, and Action,* eds. Mark Francis and Randolph T. Hester, Jr. Cambridge, Mass.: MIT Press, 1990.

"Introduction." In Edith Cobb, *The Ecology of Imagination in Childhood.* Japanese ed. Tokyo: Shishaku-sha, 1986.

"Homage to Heidegger." In *Deep Ecology,* ed. Michael Tobias. San Diego: Avant Books, 1984.

"Sociobiology and Value Systems." In *The Responsibility of the Academic Community in the Search for Absolute Values.* Vol. 2. Proceedings of the Eighth International Conference on the Unity of the Sciences. New York: International Cultural Foundation Press, 1980.

"Introduction." In *The Comedy of Survival, In Search of an Environmental Ethic,* ed. Joseph Meeker. Los Angeles: Guild of Tutors Press, 1980.

"The Conflict of Ideology and Ecology." In *The Search for Absolute Values in a Changing World.* Vol. I, Proceedings of the Sixth International Conference on the Unity of the Sciences. New York: International Cultural Foundation Press, 1977.

"Nature Study—Indoor Images, Outdoor Reality." In *Claremont Reading Conference Forty-first Yearbook,* ed. Malcolm P. Douglass. Claremont: Claremont College Press, 1977.

"Introduction." In José Ortega y Gasset, *Meditations on Hunting.* Translated by Howard B. Wescott. New York: Scribner's, 1972.

"Ecology and Man—A Viewpoint." In *The Everlasting Universe: Readings on the Ecological Revolution,* ed. Lorne J. Forstner and John H. Todd. Lexington, Mass.: Heath, 1971.

"Ecology and Man—A Viewpoint." In *It's Not Too Late,* eds. Fred Carvell and Max Tadlock. Beverly Hills: Glencoe Press, 1971.

"Ecology and Man—A Viewpoint." In *The Ecological Conscience: Values for Survival,* ed. Robert Disch. Englewood Cliffs, N.J.: Prentice-Hall, 1970.

"Ecology." In *Prophecy for the Year 2000,* ed. Irving A. Falk. New York: Julian Messner, 1970.

"The Virtues of Anonymity." In *A Reading Approach to College Writing,* ed. Martha Heasley Cox. San Francisco: Chandler, 1967.

"The Eyes Have It." In *This Is Nature: Thirty Years of the Best from Nature Magazine* ed. Richard W. Westwood. New York: Crowell, 1959.

Essays

1990s

"The Origin of Metaphor: The Animal Connection." *The Touchstone Center Journal* (1997):7–17.

"Are Pets a Healthy Link with Nature?" *CQ Researcher* Dec. 27, 1996:1145.

"Wilderness Is Where My Genome Lives." *Whole Terrain* 4(1995–1996): 12–16.

"Gaia Doubts." *American Nature Writing Newsletter* (Spring 1994):11.

"The Biological Bases of Bear Mythology and Ceremonialism." *Bears of Russia and Adjacent Countries—State of Populations.* Proceedings of the Sixth Conference of Specialists Studying Bears, Central Forest Reserve. Tver Oblast, Russia, September 6–11, 1993.

"Digging for Our Roots." *Places* (Summer 1990):68–81.

"Searching Out Kindred Spirits." *Parabola* (May 1991):86–87.

1980s

"The Philosopher, the Naturalist, and the Agony of the Planet." *The Human/Animal Connection: The Carnivore.* Vol. 8, pt 1, Sierra Nevada College Press 1 (1985):84–89.

"Celebrations of the Bear." *North American Review* (September 1985):17–25.

"The Ark of the Mind." *Parabola* (May 1983):54–59.

"On Madness and Nature." *Pitzer College Participant* (Spring 1982):9.

"Five Green Thoughts." *Massachusetts Review* (Summer 1980):273–288.

"Not Quite Fatal." Pamphlet published by the Canadian Society for Social Responsibility in Science (1980).

1970s

"Itinerant Thoughts on Place" *Pitzer College Participant* (Fall 1977):3–9.

"Place in American Culture." *North American Review* (Fall 1977):22–32.

"Ugly Is Better." *Pitzer College Participant* (Winter 1975): 10–14.

"La ecologia y el hombre." *Revista de Occidente, Ecologia y Urbanismo* (Feb.-March 1975):201–215.

"Place and Human Development." *Children, Nature, and the Urban Environment.* Proceedings of a symposium fair, May 19–23, 1975, pp. 7–12.

"Animal Rights and Human Rites." *North American Review* (Winter 1974): 35–42.

"Hunting for a Better Ecology." *North American Review* (Summer 1973): 12–15.

"Establishment and Radicals on the Environmental Crisis." *Ecology* (Aug. 1970).

"Human Ecology: Evolution and Development." *The New Natural Philosophy,* International College (1970):19.

1960s

"Whatever Happened to Human Ecology?" *BioScience* (Dec. 1967):891–894.

"The Virtues of Anonymity." *Science Review* (Sept. 17, 1966):77.

"The Wilderness as Nature." *Atlantic Naturalist* (Jan. 1965):9–14

"The Corvidean Millennium; or Letter From an Old Crow." *Perspectives in Biology and Medicine* (Spring 1964):331–342.

"The Arboreal Eye." *School Science and Mathematics* (Dec.1964):736–740.

"Aggression and the Hunt: The Tender Carnivore." *Landscape* (Fall 1964): 12–15.

"The Artist as Explorer: A Review by Paul Shepard." *Landscape* (Winter 1962–1963): 25–27.

"The Cross Valley Syndrome." *Landscape* (Spring 1961): 4–8.

"English Landscape Esthetics in the Settlement of New Zealand." *Proceedings of the AAAS Tenth Pacific Science Congress: The Role of Cultural Values in Land Use* (Aug. 31, 1961).

1950s

"A Theory of the Value of Hunting." Transactions of the Twenty-fourth North American Wildlife Conference (March, 1959): 504-512.

"Biological Perspective on the Broad Scale Use of Chemical Pesticides." *Massachusetts Audubon* 43(4)(1959):165–167.

"Reverence for Life at Lambaréné." *Landscape* (Winter 1958–1959):26–29.

"The Place of Nature in Man's World" *School Science and Mathematics* (May 1958):394–403.

"The Place of Nature in Man's World." *Atlantic Naturalist* (April 1958):85–89.

"Paintings of the New England Landscape." *College Art Journal* 17(1) (1957):30–42.

"Dead Cities in the American West." *Landscape* (Winter 1956):25–28.

"The Nature of Tourism." *Landscape* (Summer 1955):29–33.

"Montana's Marching Mountains." *Nature Magazine* (Feb. 1954):97–99.

"Something Amiss in the National Parks." *National Parks Magazine* (Oct.–Dec. 1953):150–151 and 187–190.

"They Painted What They Saw." *Landscape* (Summer 1953):6–11.

"Watching Wildlife at Crater Lake." *Audubon Magazine* (Nov. 1952):383–387.

"Experiment in Learning." *American Forests* (Oct. 1952).

"Can the Cahow Survive?" *Natural History* (Sept. 1952):303–305.

"The Dove is Doubtful Game." *Nature Magazine* (Aug.–Sept., 1952):351–352.

"Yale Conservation Club Sponsors a Series of Panels." *Connecticut Woodlands* (March 1952):9–10.

"Our Highways and Wildlife." *Nature Magazine* (Jan. 1952):34–37.

"Eyes—Clues to Life Habits." *Nature Magazine* (Nov. 1951):457–460.

"The Sportsman's Dilemma." *The Land: Notes Afield* (Summer 1951):168–171.

Acknowledgment of Sources

"The Origin of Metaphor: The Animal Connection" was originally presented as part of a lecture series for the Touchstone Center at the American Museum of Natural History in New York City in the fall of 1994 and was originally published in *Writings on the Imagination* (New York: Touchstone Center Journal, 1997), 7–17. Copyright © 1997 The Touchstone Center for Children, Inc., New York. Used by permission. The letter from "The Others" was published in Paul Shepard's *The Others: How Animals Made Us Human* (Washington, D.C.: Island Press, 1996), pp. 331–333.

"Animals and Identity Formation" (1988) and "Discoursing the Others" (1991) were originally presented by Paul Shepard at conferences of the *Journal of Curriculum Theorizing,* an interdisciplinary journal of studies on learning, teaching, and curriculum.

"The Animal: An Idea Waiting to Be Thought" (1992) is a previously unpublished essay.

"The Eyes Have It" was originally published in *This is Nature: Thirty*

Years of the Best from Nature Magazine, ed. Richard W. Westwood (New York: Crowell, 1959), pp. 193–198. It was first published in *Nature Magazine* in November 1951 as "Eyes—Clues to Life Habitats." (Good-faith efforts have been made to secure permission to reprint material from this work. If the copyright holder will contact the publisher, any necessary corrections can be made to future printings.)

"The Arboreal Eye" was originally published in *School Science and Mathematics,* (December 1964):736–40. Reprinted by permission of *School Science and Mathematics.*

"Reverence for Life at Lambaréné" was originally published in *Landscape* (Winter 1958–1959: 26–29). (Good-faith efforts have been made to secure permission to reprint material from this work. If the copyright holder will contact the publisher, any necessary corrections can be made to future printings.)

"A Theory of the Value of Hunting" was originally published in *Transactions of the Twenty-fourth North American Wildlife Conference,* March 2–4, 1959 (Washington, D.C.: Wildlife Management Institute, 1959). Reprinted by permission of The Wildlife Management Institute.

"Aggression and the Hunt: The Tender Carnivore" was originally published in *Landscape* (Fall 1964):12–15 (Good-faith efforts have been made to secure permission to reprint material from this work. If the copyright holder will contact the publisher, any necessary corrections can be made to future printings.)

"Meditations on Hunting" was originally published as the introduction to the English translation of José Ortega y Gasset's *Meditations on Hunting,* trans. Howard B. Wescott (New York: Scribner, 1972). Reprinted by permission of Scribner, a division of Simon & Schuster. Copyright © 1972, 1985 by Revista de Occidente S/A.

"The Significance of Bears" (1995) is a previously unpublished essay.

"Digging for Our Roots" was originally published in *Places* 6(4) (Summer 1990):69–8I as an accompanying essay to "Portfolio: Selection from Colstrip, Montana, Photography by David Hanson." Copyright © 1990 The Design History Foundation.

"Five Green Thoughts" was originally published in *Massachusetts Review* 21(2) (Summer 1980):273–288; it was revised by Paul Shepard in 1995.

"The Nature of Tourism" was originally published in *Landscape* (Summer 1955):29–33. (Good-faith efforts have been made to secure permission to reprint material from this work. If the copyright holder will contact the publisher, any necessary corrections can be made to future printings.)

"Whatever Happened to Human Ecology?" was originally published in *BioScience* 17(12) (December 1967):891–894. Copyright © 1967 American Institute of Biological Sciences.

"The Conflict of Ideology and Ecology" was originally published in *The Search for Absolute Values in a Changing World,* Vol. 1, Proceedings of the Sixth International Conference on the Unity of the Sciences, San Francisco, November 1977 (New York: International Cultural Foundation Press, 1977), pp. 31–40.

"Sociobiology and Value Systems" was originally published in *The Responsibility of the Academic Community in the Search for Absolute Values,* Vol. II, Proceedings of the Eighth International Conference on the Unity of the Sciences, Los Angeles, November 1979 (New York: International Cultural Foundation Press, 1980), pp. 949–955.

"Ugly Is Better" was originally published in *Pitzer College Participant* (Winter 1977):10–14.

"Itinerant Thoughts on Place" was originally published in *Pitzer College Participant* (Fall 1977):3–9.

"Place and the Child" (1990) is a previously unpublished essay.

Index